The Iron Whim

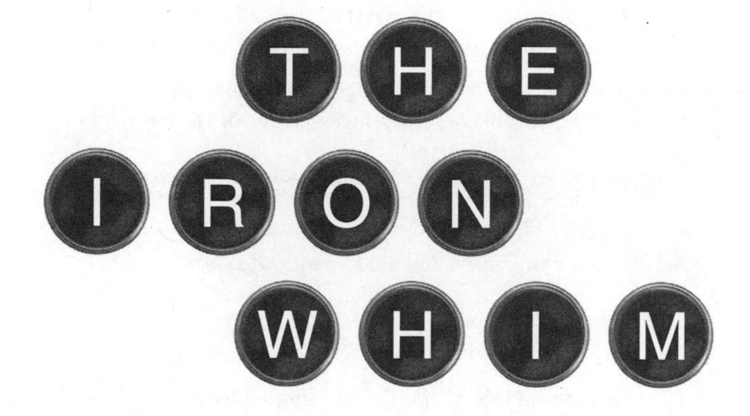

A Fragmented History of Typewriting

DARREN WERSHLER-HENRY

Library and Archives Canada Cataloguing in Publication

Wershler-Henry, Darren S. (Darren Sean), 1966–
The iron whim : a fragmented history of typewriting /
Darren Wershler-Henry.

ISBN 0-7710-8925-2

1. Typewriters – History. 2. Typewriting – History. I. Title.

Z49.A1W47 2005 652.3'009 C2005-903236-7

We acknowledge the financial support of the Government of Canada through the Book Publishing Industry Development Program and that of the Government of Ontario through the Ontario Media Development Corporation's Ontario Book Initiative. We further acknowledge the support of the Canada Council for the Arts and the Ontario Arts Council for our publishing program.

Printed and bound in Canada

This book is printed on acid-free paper that is 100% recycled, ancient-forest friendly (100% post-consumer recycled).

McClelland & Stewart Ltd.
The Canadian Publishers
75 Sherbourne Street
Toronto, Ontario
M5A 2P9
www.mcclelland.com

1 2 3 4 5 09 08 07 06 05

for Joe and Caley

It types *us*, encoding its own linear bias across the free space of the imagination.

– J. G. Ballard, "Project for a Glossary of the Twentieth Century"

Contents

The Iron Whim

Introduction

Ghosts and Machines

"More than a dozen researchers have witnessed the keys of the bizarre typewriter moving, seemingly on their own, spelling out cryptic information about the afterlife – and the consequences of evil behavior in this life."

– Ann Victoria, "Is This Old Typewriter Proof of Life After Death?", *The Weekly World News*

Typewriting is dead, but its ghosts still haunt us.

Even in our image-saturated culture, the iconic value of the typewriter looms large. Artfully grainy, sepia-toned close-up photos of its quaint circular keys grace the covers of tastefully matte-laminated paperbacks, announcing yet another volume extolling the virtues of the writing life. In magazine and billboard ads, magnified blotchy serifed fonts mimic the look of text typed on battered machines with old, dirty ribbons: pixel-perfect damaged letters that sit crookedly above or below the line with paradoxical consistency. On radio and TV, the rapid clatter of type bars hitting paper signals the beginning of news broadcasts. We all know what this sound means: important information will soon be conveyed. Typewriters may have been consigned to the dustbin of history, but their ghosts are everywhere.

What's remarkable is not that typewriting continues to haunt us, but that typewriting itself was *always* haunted.

Consider the case of Felix Pender, a successful young author of humorous stories. Pender, a character in "A Psychical Invasion,"[1] one of Algernon Blackwood's turn-of-the-century tales of the paranormal, has a problem. Though he is producing new work at an alarming rate, the young Pender is no longer capable of writing anything funny.[2] All his laughter seems "hollow and ghastly, and ideas of evil and tragedy [tread] close upon the heels of the comic."[3]

In the best Edwardian fashion, Pender's difficulties stem from his misguided attempt to learn Things Man Was Not Meant to Know. In order to experience the ludicrous in a manner that he would not normally, and therefore, presumably, to generate some new material, Pender starves himself for six hours, then takes an "experimental dose" of hashish.[4] After a slightly disconcerting laughing jag, he goes to bed, wakes late, and sits down to write:

> All that day I wrote and wrote and wrote. My sense of laughter seemed wonderfully quickened and my characters acted without effort out of the heart of true humour. I was exceedingly pleased with this result of my experiment. But when the stenographer had taken her departure and I came to read over the pages she had typed out, I recalled her sudden glances of surprise and the odd way she had looked up at me while I was dictating. I was amazed at what I read and could hardly believe I had uttered it . . .
>
> It was so distorted. The words, indeed, were mine so far as I could remember, but the meanings seemed strange. It frightened me. The sense was so altered. At the very places where my characters were intended to tickle the ribs, only curious emotions of sinister amusement resulted. Dreadful innuendoes had managed to creep into the phrases. There was laughter of a kind, but it was bizarre, horrible, distressing; and my attempt at analysis only increased my dismay. The story, as it read then, made me shudder, for by virtue of those slight changes it had come somehow to hold the soul of horror, of horror disguised as merriment.[5]

The writing so horrifies Pender that he feels compelled to destroy it. But doing so brings no relief, because he feels, both outside and within himself, "in most intimate fashion," an evil, hateful female Presence.[6]

Everything Pender has tried to write since this experience feels "as though someone else had written it." His stenographer has left him, "of course," and he has been afraid to take another.[7] Without his typewriter – at the time, the word referred to both the machine and its operator – Pender is helpless. After he takes up a pencil "in obedience to an impulse to sketch – a talent not normally mine,"[8] he finds that the face of the terrible woman he has felt an urge to draw is nothing but "lines and blots and wiggles."[9] There is no doubt about it: Pender's typewriting is haunted.

Enter Blackwood's hero: John Silence, Physician Extraordinary – a sort of Sherlock Holmes of the paranormal. A wealthy philanthropist, Silence demands no remuneration for his services. Further, Silence has a highly specialized practice: he only takes on cases from the "very large class of ill-paid, self-respecting workers, often followers of the arts" suffering from maladies "of that intangible, elusive, and difficult nature best described as psychical afflictions."[10] With the help of his faithful dog and cat (who are "almost continuously conscious of a larger field of visions, too detailed even for a photographic camera, and quite beyond the reach of normal human organs"[11]), and the knowledge gleaned during his many years of training and research into the occult, Silence exorcises the Presences (it turns out there are more than one), much to Pender's relief.

With the help of his *own* psychic typewriter, Silence is even able to definitively establish the identity of the chief Presence for Pender: "Dr. Silence drew a typewritten paper from his pocket. 'I can satisfy you to some extent,' he said, running his eye over the sheets, and then replacing them in his coat; 'for by my secretary's investigations I have been able to check certain information obtained in the hypnotic trance by a "sensitive" who helps me in such cases.'"[12]

The ghost haunting Pender was that of "a woman of singularly atrocious life and character who finally suffered death by hanging, after a series of crimes that appalled the whole of England," a practitioner of "the resources of the lower magic" who once lived in a now-demolished house built on the same property as Pender's.[13] The true identity of story tells us about the implicit rules that have structured our perceptions of typewriting from the moment of its inception.

"All perception, as you know, is the result of vibrations," Silence tells Pender early in their acquaintance. Psychics, dogs, and cats merely perceive a wider spectrum of those vibrations than the bulk of humanity.[14] In some cases, mind-altering drugs or violent emotions may also open someone's "inner being to a cognisance of the astral region."[15] Once a brain has been "suitably prepared" in such a manner, "strong thoughts and purposes," whether they originate with a dead person or some other denizen of the astral realms, can alter or replace the signals the brain would normally receive.[16] The era into which typewriting was born was swayed almost equally by the discourses of science and mysticism. From its perspective, drugs are as much a spiritual tuning device as they are a medical technology, and ghosts, it turns out, are nothing more than ectoplasmic ham-radio operators.

In a world where all perception is vibration, a writer is someone who can attune themselves to particular kinds of vibrations and refine them into text. By Blackwood's era, the typewriter was an essential part of that process, a device that aided and abetted the reception and processing of inspirational vibrations for the writer. No *real* writer could possibly continue to make use of a pencil when a typewriter was available.

Moreover, the "typewriter" that the author employs may not be a mechanical device. There is almost always an amanuensis or typist taking dictation from the writer and operating the machine, and that amanuensis is almost always a woman. That the ghost (which at one point multiplies into a whole houseful of ghosts) interfering with Pender's sense of humour is not only female but "evil" suggests that the "good" or proper place for a woman is as the passive interface for the typewriting assemblage, not as the source of creativity. The amanuensis, however, may not always stick around to tolerate such restrictive assumptions; for a woman who can type, opportunities for employment abound elsewhere.

While the entire typewriting assemblage – inspirational voice or voices (originating both within and without the dictator), dictator, machine, and amanuensis – is capricious and subject to hijackings and rebellions of various sorts, it is also, inevitably, the source of the truth behind events. Yes, there is a contradiction here: just as surely as Pender's typewriting produces lies, John Silence's typewriting produces truth. This contradiction persists throughout typewriting history. Over and over again, mass media presents typewriters as the pre-eminent modern symbol of written truth, but just as often, typewritten texts turn out to be capricious, riddled with errors, and occasionally outright *wrong*. That doesn't seem to bother anyone in the slightest, because our memories of typewriting are clouded by nostalgia.

Nostalgia for typewriting is everywhere. Once, typewriting symbolized all that was antithetical to poetry; it was cold, mechanical, awkward. Now, however, through the misty lens of nostalgia created by several centuries of typewriting's own propaganda, we believe that typewriting *is* poetry: precise, clean, elegant in its minimalism. In January of 2005, while flipping through *Now*, a weekly Toronto entertainment paper, I found an ad for the Ryerson University writing workshops at the G. Raymond Chang School of Continuing Education, prominently featuring a close-up of a pristine antique manual typewriter. "Feel a poem bursting forth or a novel taking over your life?"[17] asks the accompanying text. Yes, but . . . in a world where writing technology is, indisputably, dominated by computers and has been for at least two decades, why is the typewriter the vehicle mediating that "bursting forth"? How can I be sure it's *my* poem bursting forth through my typewriter, *my* novel taking over my life? How do I know that it's not actually a novel or poem belonging to the ghost of a woman of singularly atrocious life and character who finally suffered death by hanging, after a series of crimes that appalled the whole of England? Or something even stranger? And, if I decide to try to type out that novel or poem, what sorts of horrible and eldritch technologies will I be subjected to? Will typewriting *change* me, adapt me, body and soul, to its own ends?

Of course it will. And I have no way of knowing where those texts that typewriting "inspires" really come from. Finding the answers to these questions requires a kind of arcane cross between forensic science and archaeology: all that's left of the logic of typewriting now is fragments. Typewriting died a violent death, and any Physician Extraordinary worth his salt will tell you that violent deaths lead to hauntings. Exorcising its ghosts requires a willingness to poke around in some of the stranger corners of writing from over the past two centuries. And, as with "A Psychical Invasion," there are no real surprises at the story's end. We know from the beginning what typewriting produces. What we can learn along the way is how the rules that produce typewriting function, and the kinds of things that typewriting will allow us to say. The French technology theorist Paul Virilio claims that in a contemporary setting, our dilemmas always involve a need to discern what, exactly, are the problems we are facing, not to find answers for them. In such circumstances, the job of the writer is to "exhibit the accident"[18] . . . which is exactly what I'm about to do.

Part

Archaeology: Beginning at the End

The way in which objects are thrown down in the streets reveals a certain contempt, and it may be witnessed often enough in that singular, abrupt gesture in which people toss something down on the pavement, or let it fly in a little arc from a car window.

– Julian Stallabrass, "Trash"

Chapter 1

Conclusion: Royal Road Test

Sunday, August 21, 1966: a perfect day for an execution.

The surfeit of detail is almost obscene. Approximately 122 miles southwest of Las Vegas, a 1963 Buick Le Sabre, licence plate FUP 744, is hammering along U.S. Highway 91 (Interstate 15). At 5:07 p.m., the passenger window rolls down, and a Royal Model X typewriter hits the pavement at 90 miles per hour.

The wreckage stretches along 189 feet of asphalt and Nevada desert . . . and back in time over 252 years.

The typewriter itself was only ever a symbol of something much larger. The fragments that lie glittering under the desert sun are the shattered remains of a system that dictated many aspects of not only how Western cultures performed the act of writing, but also of how we organized work, play, and even the education and production of "useful" members of society, for two and a half centuries.

Forensics

Three men are responsible for the death of the typewriter: Ed Ruscha (the driver and artist behind the project), Mason Williams (the thrower, and a writer; Ruscha was insistent that an actual

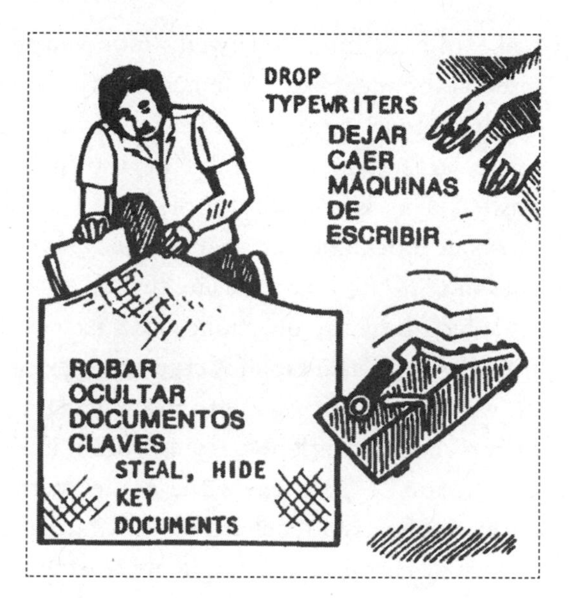

According to the CIA, dropping your employer's typewriters is still a subversive act.

writer had to deal the death-blow to the machine), and Patrick Blackwell (the photographer). In the final image of *Royal Road Test*,[1] the artists' book that documents this event, their shadows loom over the broken wreckage of the typewriter like those of Quentin Tarantino's *Reservoir Dogs*,[2] distorted, gesticulating, very possibly mocking the broken frame of their victim.

As an object, *Royal Road Test* is a paragon of ironic design: a small, canary-yellow spiral-bound book resembling nothing so much as a manual of operating instructions for the machine whose destruction it documents. Its contents are primarily photographic, not textual; forensic, but also oddly didactic – a how-to manual for potential artist-Luddites. The book is a cross between a pulp fiction *fotonovela* and the pamphlets that the CIA drop into communist nations, providing instructions about how to bring down the regime by performing actions such as dropping typewriters to damage them.[3]

As the reader flips through the pages of *Royal Road Test*, the artists morph from executioners to detectives, combing the crash scene and labelling Blackwell's monochrome photos of the scattered components of the machine with pseudo-scientific rigour: "Rubber Twirler Knob"; "Piece of Back Cover Acoustical Padding (heavy green paper)"; "Tab Key Top (photographed as found in bush)." This is a reality book before reality TV, presenting the artists as mockumentarian/archaeologists in the vein of the detectives of *CSI: Las Vegas*, teasing tiny clues out of the debris.

Which begs the question: what, exactly, are they looking for? Even the book's epigraph is cryptic, mock-heroic, Chandleresque – "It was too directly bound to its own anguish to be anything other than a cry of negation; carrying within itself, the seeds of its own destruction" – and leaves the reader with more questions than answers. How is a typewriter a cry of negation? What anguish can a machine carry? And what, if any, are the seeds of destruction that it bore within itself?

Wreckage Mirage

From the perspective of the twenty-first century, almost forty years after the event, it's tempting to read *Royal Road Test* as a tragedy or, at least, a melodrama. The book's sombre black-and-white photographs detail the spectacular end of a machine tested to the point of destruction. And not just any machine, but the typewriter – a machine for which people harbour a particular and unusual sense of nostalgia.

In the American collective imagination, the Nevada desert is the scene of the classic "mob hit" (see *Goodfellas* and the novels of James Ellroy for starters). But more than fictional characters meet their end in the desert. We're also used to seeing many of the mechanical icons of popular culture reduced to a handful of jagged debris scattered across the American southwest: James Dean's 1955 Porsche Spyder, torn nearly in two on the highway between L.A. and Salinas; the space shuttle *Columbia*, strewn from orbit across the fields of central Texas in 2003. However, time and the desert

Ed Ruscha (left) and Mason Williams, surveying the wreckage of the Royal Model X.

play tricks around a ruin. Both nostalgia and nihilism are mirages, projections of our own desires onto the haze and heat. To see beyond those mirages requires a slow sifting through fragments scattered over time and across space.

The site of this survey is vast, because the fragmented archive of typewriting extends back at least as far as the eighteenth century. And, like many crime scenes and archaeological sites, the record it presents is incomplete, containing various and sundry dead ends, false starts, freaks, curios, and oddities. But from these fragments and partial objects it is possible to re-collect not only the history of a machine, but also to describe the logic it enforced: a logic that structured almost two centuries of writing.

There are plenty of books detailing the history of the typewriter as a machine; the most accurate and authoritative of these is

Michael H. Adler's *The Writing Machine* (London: George Allen & Unwin, 1973). *The Iron Whim* is different. The crime scene investigators who picked up the pieces of James Dean's Porsche, and the scientists who gathered the fragments of the *Columbia*, weren't particularly interested in the machines as machines. They were interested in using the fragments to construct a story about how and why a particular event occurred. Discovering how the parts of a machine work together – and how they fail to work at crucial moments – points to the underlying systems that brought those machines into being in the first place and determined the conditions under which they operated. Likewise, I'm interested in typewriting as a *discourse*: one of the systems of ideas and rules that structure our lives in ways that are subtle and brutal by turns. My goal in writing this book is to begin to understand how typewriting shaped and changed not only literature, but also our culture, and even our sense of ourselves.

Meeting this goal requires a special kind of archaeology. British art historian Julian Stallabrass notes that during the brief period when something exists as trash, its material nature becomes visible. In other words, it's possible to perceive not only the work that went into its construction, but also the rules and systems that governed that work, however arbitrary and weird those rules and systems might be. When an object is new, we see only the gloss and shine; when it's broken and faded, we see the arbitrariness and carelessness of its construction, the blatant pitches to values we have long since abandoned, the naked avarice of the manufacturer. In order to see a manufactured object clearly, you have to break it and throw it away.[4]

Throwing a typewriter out of the window of a moving car, then, is part of the process of discovering the secret rules that governed our use of the machine. The artistic process of framing, photographing, and arranging the resulting images creates a narrative that we can read and analyze: an allegory for the way that contemporary capital operates.[5] *Royal Road Test* reveals each constituent component of the typewriter with the specificity of an operator's manual.

After the crash, the typewriter looks very different: not the triumphant culmination of science and industry's steady march toward a rational future where machines solve our every problem, but a loose and messy aggregate of parts that is subject to failure, error, and malfunction. Stallabrass elegantly summarizes: "Torn, dirtied, or broken, thrown into combination with other fragmentary objects, while it remains itself, it becomes a broken shell, its meaning reaching out to its partners in a forlorn but telling narrative."[6] It is that narrative – let's call it "typewriting" – gathered from some of the decidedly odd corners of Western writing – that this book details.

Chapter 2

The Archive of Typewriting

For much of the last two centuries, typewriting *was* writing. Its logic shaped not only written documents, but also bodies, workplaces and practices, institutions and politics. We are barely outside of it, and only now capable of describing it by sifting through the archive of documents by and about typewriters for information about the rules that determined its operations.

The rules of the systems that govern our lives are always at least partly invisible to us, at least while they're still functioning. Philosopher Michel Foucault writes that "it is not possible to describe our own archive, since it is from within these rules that we speak, since it is that which gives to what we can say." Moreover, when we look back at the archives that we *can* see – the systems of past rules and principles of order, especially those that have become recently visible, like the archive of typewriting – we are incapable of describing them in their totality. The archive is the sum total of the "things we can no longer say." Over the passage of time, as we move further and further away from a mode of thinking, bits and pieces become visible, suggesting the vague outlines of overall patterns and principles.[1]

Royal Road Test dramatizes the slow emergence of the archive of typewriting. As each knobby and irreducible component appears, we assign it a name, a value, and an approximate function, but this process doesn't really create any certainty or value so much as it *pretends* to.

Shamanic Induction Devices

There's no quicker way to verify that the typewriter fragments people have sifted out of the rubble are highly marketable than by mounting a surfing expedition to eBay. Not only is eBay one of the Internet's killer apps – a website with an actual working business model – it's also a rich source of information that speaks volumes about the archives of our recent past.

With characteristically dense prose, science fiction writer William Gibson dubs eBay a "Babelian Object-Library" and claims to use it as a "shamanic induction device": a method of generating new ideas by sifting through its bazaar of juxtaposed cultural fragments in much the same way that I had to sift through the fragments of typewriting to assemble this book.[2] Gibson, a pre-eminent *bricoleur* in our culture of *bricoleurs*, is drawn to eBay because of his own past. During the 1970s, he made a living as a "picker," painstakingly combing thrift shops and Sally Anns for (paraphrasing another great *bricoleur*, Luis Buñuel) "objects of obscure desire" that he knew he could resell up-market to specialist dealers, and turn a small profit. Gibson sees nostalgia as a new religion; unable to locate it elsewhere, the aging baby boomer generation now searches for meaning in inanimate objects. This search for meaning in the relics of the past is itself what creates their value. We have become a culture of amateur curators, where everyone is able to build meaning by buying and organizing someone else's trash.[3]

In *We Want Some Too: Underground Desire and the Reinvention of Mass Culture*, Hal Niedzviecki begins to work toward providing a rationale for the surge in amateur curation, including the collection and cataloguing of objects that seem totally devoid of value. For Niedzviecki, the establishment of a personal museum dedicated to some small segment of the flow of mass cultural detritus is an attempt through participation to avoid alienation – and to insert some form of meaningful narrative back into the flow. By trading scraps of information and other forms of mass cultural residue as though they had value, Niedzviecki suggests that they *become* valuable.[4] Collecting weird crap like Smurf lunchboxes and antique video game consoles may be ridiculous, but actually admitting that would leave us with *no* mechanism for constructing meaning. Niedzviecki describes this state as "triumphantly sad,"[5] because while our gimcrack museums announce the seeming impossibility of escape from mass culture, they simultaneously allow us to address our all-consuming problem – which he contends is the utter absence of belief in the value of our daily existence.[6]

A Typewriter Reliquary

In a post-*Royal Road Test* world, old typewriters and typewriter fragments are cheap and available in bewildering variety.

On a crisp October afternoon – in every respect, a typical eBay day – after entering the single word "typewriter" in the site's search window, I find twenty-five pages of items on auction related to typewriters.[7] Many of the descriptions of these items make prominent use of the word "vintage": "vintage royal typewriter, beveled glass, 1920s"; "Vintage 1913 Oliver Typewriter #9 Unusual"; "Vintage Royal Typewriter In case never used"; "Vintage Marx Toy Typewriter Marxwriter w/box!" (all *sic*). Through the magical application of one adjective, apparently anything can be equated to a fine wine, and therefore rendered worthy of the attention of a connoisseur . . . or, conversely, anyone interested in such "vintage" objects must necessarily *be* a connoisseur, even if the adjective has become nearly meaningless.

The sheer variety of typewriter-related detritus up for auction on eBay is staggering. It's as though Gibsonian pickers have been walking the highways of the nation, collecting the broken wreckage that Ruscha, Williams, and Blackwell missed: "Vintage Miller Elk Typewriter Ribbon Tin NR!"; "Vintage Smith Premier Typewriter Oil Bottle"; "Lot of 5 IBM Typewriter Font Balls"; "Vintage Remington Typewriter Keys 48 Fl"; "SCM Typewriter repair parts 40 pieces"; "SET OF 48 TYPEWRITER KEYS REMINGTON CELULOID"; "OLD Fabric UNDERWOOD Typewriter Ribbon~MIB"; "RARE – 'BATTLESHIP BRAND' typewriter ribbon tin."

As with the relics of saints and pop stars, sometimes the parts are worth more than the whole. Fragments of the typewriters of yesteryear often cost more than the numerous functioning electric business machines from the 1970s and 1980s, forgotten and forlorn as their users abandoned them for cheap desktop computers. Of these machines, the only ones that rise above the fifty-dollar mark are those that have some sort of link to the notion of authorship. On this particular day, writer/veterinarian James Herriot's typewriter – a black Brother Super 7800 electric typewriter (1961–1980) – can be yours for a mere £122.

There's even typewriter jewellery, and the supplies for making it: "Vintage Typewriter Key Keys Costume Earrings"; "Vintage Typewriter Key Keys Tie Clip & Tack"; "Vintage Typewriter Key Keys Cuff Bracelet WOW"; "ANTIQUE Typewriter Keys Key Bracelet – RANDOM"; "BRACELET FORMS FOR TYPEWRITER KEYS." With its stained ivory-coloured Bakelite keys and carious black-enamelled lettering, the bracelets have the grotesque charm of Victorian baubles fashioned from the baby teeth and bones of the recently departed.

There are messages spelled out on those bracelets as well – messages for us about the gender of the majority of typists, and the attitude of their employers (those who dictated the words being typed, and perhaps even those who bestowed the bracelets).

"Old Typewriter Key Bracelet . . . 'GAL FRIDAY+'" is typical; "Old Typewriter Key Bracelet . . . '+++BINGO+++'" suggests that "Gal Friday" might have spent her nights away from the typewriter with a bingo dauber in one hand and a Marlboro Light in the other.

There are representations of the typewriter for sale as well as actual relics. The force of nostalgia has recreated the typewriter as a bewildering array of kitschy ornaments: "SWAROVSKI CRYSTAL MEMORIES TYPEWRITER"; "Mini Clock: Silver Typewriter Miniature-cool!"; "VTG 14K GOLD MECHANICAL RUBY TYPEWRITER CHARM"; "1960s Pink Typewriter Figural Ashtray GREAT"; "TYPEWRITER & INK BOTTLE SALT AND PEPPER SHAKERS"; "Limoges Box Typewriter France HP NR"; "2 Old Chocolate Molds: ACCORDION & TYPE WRITER"; and so on. These tiny mementoes are more than mere adornment; they also suggest that typewriting had its own form of heroism (or militarism). How else can we explain the "1921 Speed & Accuracy medal from the Underwood Typewriter co." or the "REMINGTON MEDAL OF AWARD TYPEWRITER PIN-STERLING SILVER" and the "Vintage Remington Typewriter Award Silver Pin"?

And then there are the documents. eBay is littered with postcards, posters, photos, and old advertisements clipped from magazines and newspapers, depicting typing machines and extolling their myriad virtues. Postcard collectors might be interested in "i23 Syracuse NY Smith Typewriter Wks Postcard," "Greta Garbo at her typewriter," or Lady Liberty at *her* typewriter.

Some of these documents point to various key moments in the machine's history: "patent specifications dated MARCH 19th, 1872 (NOT A REPRODUCTION) – from RASMUS MALLING HANSEN, CHAPLAIN TO THE DEAF AND DUMB INSTITUTE and C. P. JURGENSEN, MATHEMATICAL INSTRUMENT MAKER." The Patent Museum, an online retailer, even sells a replica of Christopher Latham Sholes's U.S. patent for the typewriter: "A drawing specification sheet from the patent has been reproduced on parchment paper, secured to a quality 8" × 10" Matboard, and is ready to install into a frame." A diploma reveals that it was once possible to hold a "Doctorate of Typewriters," though the awarding institution and the value of the diploma remain vague. Was this a diploma for typewriter repair? Typewriter scholarship? Typewriter use? And how effective was it in securing employment for its holder? As insignificant as they may seem, these documents serve an important function. They keep the typewriter highly visible at a moment when it is on the verge of disappearing for good.

If scavenging eBay reveals that it's possible to be indoctrinated into knowledge of the typewriter, it also reveals that the process of indoctrination begins early. Consider the following toys up for auction: a Marx Deluxe dial typewriter from the 1930s; "VINTAGE TOY SIMPLEX TYPEWRITER W/BOX – 1934"; "American-Flyer Typewriter, U.S.Pat. 1-907-379-Made In U.S.A."; "OLD UNIQUE ART TIN LITHO TOY TYPEWRITER"; "Corgi Commander Typewriter. Described on the box as 'The Typewriter made Specially For Boys'"; "'BARBIE' TYPEWRITER & BARBIE BOOK"; the "Cub Reporter Tin Typewriter"; "Child's Typewriter Tom Thumb Western Stamping." Boys who use typewriters are cub reporters, commanders, cowboys; girls are relegated to follow in Barbie's footsteps. Can the GAL FRIDAY bracelet be far behind?

Chapter 3

Typewriter Nostalgia

What, exactly, do the people who pay for old typewriter ribbon tins and other "vintage" objects think that they're buying?

Joshua Wolf Shenk's article "The Things We Carry" offers some suggestions.[1] It's a meditation on the phenomenal popularity of *Antiques Roadshow* – specifically, on the way the show and its various imitators convert the narratives of our lives (and the artifacts that pass through them) into monetary value.

Shenk's exploration of the commodification of emotion begins with his grandfather's Hebrew typewriter. Long past the point where it was in daily use, the typewriter has become the relic at the heart of Shenk's secular shrine to his ancestors: "In my apartment, as in my imagination, Rabbi Wolf's typewriter commands a place of honor. It sits now next to my desk, its open top covered by a small silk handkerchief."[2] It is, in other words, a kind of family snapshot, draped in crepe to signify a death. Philosopher Jean Baudrillard uses exactly this language to describe the function of an antique: like the portrait of a dead relative, it is a sort of monument whose purpose amounts to an impossible attempt to suppress time by providing a summation of a complex being that can't help but fall short of being definitive.[3]

Shenk sees all too clearly the various contradictions that are embodied in his attitude toward this machine. "I never use it – the ribbon is dry, the platen so hard that it could crack the type. But even if I replaced the ribbon and platen, and oiled the gears, I could still only play at the typewriter as a toddler might a piano, for I know Hebrew about as well as a young child knows scales."[4] He doesn't use it; he *owns* it . . . but this doesn't mean that the machine has stopped functioning powerfully on the level of imagination.

While many antiques no longer fulfil the practical tasks they once performed, they all do one thing in common – they serve as a sign of the passage of time.[5] Functioning objects are of the present, but antiques, whose function is obsolete, bear witness to the elapsing of an indefinite amount of time. There is always something false about even the finest of antiques, because their existence is paradoxical. Antiques represent genuineness in an abstract system of exchange where all value is arbitrary.[6] One person's trash is another's treasure, which is why objects from as recently as the 1970s and 1980s can be "antique" in this sense.

The ghosts that Shenk's grandfather's typewriter evokes are the embodiment of pastness: not memories, but fantasies. "If I strain, I can picture my grandfather hunched over it, striking its keys," writes Shenk. "But I am touching an absence. The typewriter speaks of my grandfather but is, for me, forever silent. It is potentially warm but perpetually cold."[7] This is not necessarily a bad thing; for Shenk, it's the interplay of connection and disconnection, contact and loss, that creates value. Like the typewriter in *Royal Road Test*, the rabbi's typewriter is simultaneously lost and found.

Despite these high-minded and philosophical sentiments, Shenk still feels compelled to assign a dollar value to his sentiment as a kind of attempt to determine *exactly* how "authentic" this typewriter is. Shenk's quest for lost time is rewarded, on visiting Martin Tytell's legendary (well, legendary among people who care about typewriters) typewriter shop in Manhattan, with the information that his grandfather's typewriter is "*very* collectible" and "worth twenty-five hundred dollars." "The rule of thumb with typewriters . . . is that if it doesn't look like what you and I perceive as a typewriter, it's worth money."[8] However, another appraiser tells Shenk that the typewriter is worth more like one hundred to one hundred and fifty dollars.[9]

The exception to this rule is if there is a strong connection between the machine and a celebrity. Attaching a famous name to a machine is an easier way to create value than is addressing the collector's obsessive need to know all of the details (What model is it? When was it made? Where? By whom? How many of them are there? Are there variants?) that authenticate an antique.[10] Ian Fleming's Remington typewriter fetched more than fifty-five thousand pounds at auction[11] (a far cry from what poor old James Herriot's machine was worth). Of course, Fleming's machine was plated in gold . . .

The end of Shenk's story is almost like a rabbinical parable. The contrite author concludes that value is "not a typewriter but what it can conjure,"[12] its invisible connections, what *it* ushers forth . . . And for some people, typewriters conjure a great deal.

Pastness

Why do people persist with all of this picking, gutter sniping, and scavenging for pieces of the typewriter? The answer, in a word, is nostalgia.

As the typewriter relics on eBay demonstrate clearly, nostalgia is a major element of our economy, because the past can be and is easily and endlessly repackaged and sold. In his monolithic tome *Postmodernism*, Fredric Jameson refers to this as the feeling of "pastness,"[13] in order to distinguish it from the reality of actually living in the past moment. Jameson contends that what people are actually longing for is not the object itself in its original context, but

Typewriter kitsch and nostalgia overflow in this idealized image of the Dionne quintuplets learning to type.

the comforting memory of memory, the nostalgia of nostalgia, of a time when things – in this case, writing – "meant something" and it was possible to ask the big questions without wondering about the presumptions involved.[14]

The effect of nostalgia on our perception of the past is considerable. It's like a thick smear of Vaseline on the lens of a movie camera, blurring our objectivity. From the far side of the millennial divide, a photo of a typewriter doesn't just show a machine but an icon of unalienated modernist writing. The typewriter has become the symbol of a non-existent sepia-toned era when people typed passionately late into the night under the flickering light of a single naked bulb, sleeves rolled up, suspenders hanging down, lighting each new cigarette off the smouldering butt of the last, occasionally taking a pull from the bottle of bourbon in the bottom drawer of the filing cabinet.

Humorist David Sedaris simultaneously personifies and parodies this attitude in his story "Nutcracker.com."[15] Sedaris writes, "I hate computers for any number of reasons, but I despise them most for what they've done to my friend the typewriter. In a democratic country you'd think there would be room for both of them, but computers won't rest until I'm making my ribbons from torn shirts and brewing Wite-Out in my bathtub. Their goal is to place the IBM Selectric II beside the feather quill and chisel in the museum of antiquated writing instruments."[16] Sedaris contends that it is the *illusion* of work that the typewriter creates, through wasted resources and noise, that makes it a compelling tool for holdout writers. "Unlike the faint scurry raised by fingers against a plastic computer keyboard, the smack and clatter of a typewriter suggests that you're actually building something. At the end of a miserable day, instead of grieving my virtual nothing, I can always look at my loaded wastebasket and tell myself that if I failed, at least I took a few trees down with me."[17]

Here's an object lesson: walk down to the corner bookstore and see how many books – novels, memoirs, and anthologies alike – sport covers featuring grainy sepia-toned close-up photos of typewriter keyboards. You'll be there a while, I guarantee it, because there are far too many examples to bother citing. The typewriter is *the* pre-eminent symbol for earnest, unalienated writing and one of the biggest visual clichés of our age.

Think for a moment about how nostalgia functions. Signs and objects (the typewriter included) persist to some degree even after the era in which they originated has long since crumbled into dust. In later times, those signs and objects are assigned new values and meanings, some positive, some negative. Typewriting does not mean the same thing now that it did even fifty years ago. Baudrillard makes an important observation: the sense of authenticity that antiques produce is more mythological than historical. Our world has nothing to do with authenticity, Baudrillard says, yet antiques continue to insist on their own authenticity.[18] We play along because antiques are part of our desperate attempts to either stop or reverse time.[19]

Antiques act as a kind of emotional and intellectual life preserver; we expect them to lift us out of the sea of our present uncertainties and surround us with the reassuring comfort of the known.

The End of Journalism?

Aside from clichéd cover art, how does typewriter nostalgia affect us in the present? Often, it provokes strong emotional reactions that are rarely rational.

In his article "The Weight of Words,"[20] Toronto journalist Stephen Knight relates the story of his rescue of a rusty, broken Underwood typewriter from the curbside (Torontonians signal that stuff is up for grabs by placing it out by the sidewalk. Inevitably, within minutes, someone will have discovered it and carted it off).

When he stops to mull over his actions, Knight embarks on an exercise in qualified nostalgia. He can't fully explain why he stops to pick up the machine "even though it seemed slightly precious and a bit embarrassing" because "crusty hacks everywhere claim to have one, and their frequent, florid claims of undying love for the old beasts have always seemed to me the worst kind of mawkish nostalgia."

If the gesture seems so mawkish and clichéd, then why perform it at all? Because, as a writer, Knight believes that words have intrinsic value, and because it's the typewriter that actually produces the words, it must also be valuable. He actually goes a step further, positioning the machine as a kind of literary Ark of the Covenant: "Never has there been a better physical manifestation of this truth than the bulky, metal contraption that threatened to lengthen my arms and buckle my knees in the few steps from the curb to the car." Lugging a computer around would never be so arduous, but then no one has ever positioned the computer as a symbol of unalienated labour. The physical effort that's required to salvage the machine only strengthens Knight's conviction that it's somehow connected to actual work, and therefore to real value.

Knight's article is characterized by a longing for a kind of journalistic Camelot – the pastness of his profession. Words, he writes, "when used economically and well," become a sign for political integrity and democracy. Presumably, the mythical time when words had unequivocal meaning, and the machines used to produce them were free from error or duplicity, was the antithesis of the present, a world of information flows and endless cheap digitized text.

It's not surprising that many journalists feel threatened by computers. Their entire profession is under siege from weblogging and other forms of networked writing made possible by computers. One important function of the antique is to provide us with a kind of alibi for the inauthenticity of our existence.[21] However, while owning a typewriter may make a writer feel removed from the endless cycle of new computer product and instant obsolescence, all it really does is demonstrate that the writer experiences the immediacy of the computerized and networked world as inauthentic and unsatisfying.[22]

So what does Knight do with his newly rescued vessel of truth? Does he use it to write something, anything – even his column? No. He puts it down in his basement with his hockey gear. The mention of the sporting equipment is an interesting detail. While Knight's hockey gear surely also bears a whiff of childhood nostalgia (among more pungent odours), at least it has a chance of being brought out into the light and used on occasion. Not only does the old Underwood fail to perform its original function, it is ineffectual as a kind of ancestral totem . . . or even a memento mori.

I ought to know. The poet Steve McCaffery's typewriter sits at the top of my staircase like a statue in a shrine or a family photo on a piano. I'd never *dream* of trying to write a book on a typewriter. I'm an unabashed techno-geek, and I write on a machine whose power and elegance make the bridge computers on the average TV starship look like they're cobbled together from old walkie-talkies, LCD fish locators, and spray-painted Barbie purses (they are, but that's a subject for another book) . . . and I'm already wondering about getting something speedier. But I do occasionally stare at the typewriter for a few seconds when I go up or down the stairs, if only to convince myself that I am extremely lucky not to have to use it.

Curmudgeons

One of the more interesting features of nostalgia is the manner in which it co-opts its critics into participating in the extension of the very thing they wish to counter. Writing about nostalgia for the typewriter produces, among other things, nostalgia for the typewriter.

One of the most nostalgic typewriter books is also one of the most recent: Paul Auster's *The Story of My Typewriter*, with oil paintings by artist Sam Messer.[23] Physically, the book utilizes all the features of a children's book – squarish trim size, large type, and copious full-colour illustrations throughout.

The Story of My Typewriter begins, like this book, with the destruction of a typewriter. Auster writes, "It was July 1974, and when I unpacked my bags I discovered that my little Hermes typewriter had been destroyed. The cover was smashed in, the keys were mangled and twisted out of shape, and there was no hope of ever having it repaired."[24]

Auster relies on the comfortable, slightly fusty language of the literary curmudgeon to defend his use of the manual typewriter. When he refers to electric typewriters at all, it is with the epithet "those contraptions" (you can almost hear him banging his cane on the porch for emphasis), which he claims to avoid because of their troubling signs of life: "the jitterbug pulse of alternating current in my fingers."[25] Predictably, Auster's take on computers ("those marvels"[26]) is similar. As he listens to the "horror stories" about what happens when one pushes a wrong button on a computer, he complains, "I knew that if there was a wrong button to be pushed, I would eventually push it."[27] This is particularly interesting because in the very passage where Auster mentions his horror of pushing a wrong button, a typo appears in what is presumably a text produced on his manual typewriter: "My friends made fun of me for resisiting [*sic*] the new ways."[28] Though this error could have entered into the text at the typesetting stage, the net effect remains destabilizing. Despite any wishes to the contrary, a typewritten text must be mediated through a computer at some point if it is going to appear in a professionally published book.

Auster's description of his quiet manual typewriter, an Olympia that presumably succeeded the smashed Hermes, as a fragile sentient being enables the reader to empathize with it even though it is an inanimate object. When Auster's infant son snaps off the carriage return arm, the incident takes on the quality of a playpen mishap where the presiding adult desperately tries not to play favourites: "That wasn't the typewriter's fault."[29] After having the

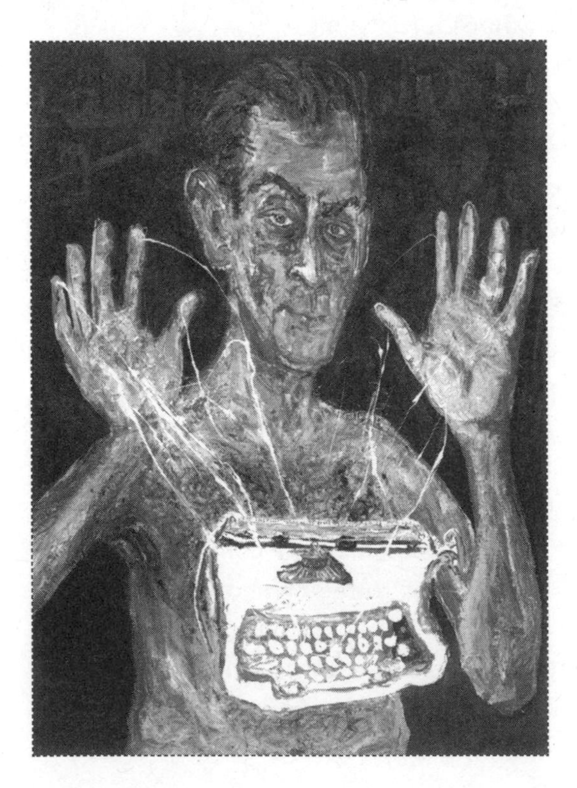

In Sam Messer's portrait of Paul Auster and his typewriter, which is the puppet and which the puppeteer?

arm soldered back in place, Auster adopts the tone of the relieved parent: "There is a small scar on that spot now, but the operation was a success."[30] When Auster's friends start to use computers, his anxiety begins to blossom. He refers to his typewriter as "an endangered species,"[31] and drops his curmudgeonly demeanour long enough to admit he is becoming fond of the machine.[32]

In a classic example of displacement, it's actually the artist Sam Messer whom Auster accuses of anthropomorphizing his typewriter. Messer, a friend and frequent guest of Auster's, becomes obsessed with the typewriter soon after his first visit to Auster's home. Auster writes that on Messer's second visit Messer "asked my wife if he could go downstairs to my work room and have another look at the typewriter. God knows what he did down there, but I have never doubted that the typewriter spoke to him. In due course, I believe he has even managed to persuade it to bare its soul."[33]

Auster's attribution of a soul to his Olympia is not an accidental or isolated instant of animism. It's common for writers not only to attribute souls to their machines, but to attribute *someone or something else's soul* to their machine. In other words, most typewriters are possessed. Auster's is no exception: "Sam has taken possession of my typewriter, and little by little he has turned an inanimate object into a being with a personality and a presence in the world. The typewriter has moods and desires now, it expresses dark angers and exuberant joys, and trapped within its gray, metallic body, you would almost swear that you could hear the beating of a heart."[34] There is a subtle but important shift occurring. As the story proceeds, we learn that the typewriter was showing signs of life long before Auster's claim that Messer imbued it with personality. When Messer first visits Auster's house, the typewriter speaks to him *before* he paints a picture of it.[35]

So the typewriter is possessed, but do the spirits possessing it play a role in the creation of the art? Here are the last lines of the book: "The typewriter is on the kitchen table, and my hands are on the typewriter. Letter by letter, I have watched it write these words."[36] Who or what is doing the writing? The image facing the last page depicts a man (presumably Auster) with lines of energy connecting his fingers to the typewriter keys. The image suggests a master/puppet relationship, but is Auster or the typewriter the puppet?

Perhaps the idea of a two-party relationship in typewriting is itself a categorical mistake. After all, there are at least four possible entities calling the shots behind the production of this little book: Auster, Messer, the Olympia, and whatever "spirits" might be possessing the machine. Perhaps it's necessary to reconceptualize the problem entirely. In typewriting, authorship is always a collective enterprise. The voice (or voices) doing the dictating, the person (or persons) receiving the dictation and doing the typing, and the typing machine (or machines) operating as an *assemblage* that produces text.

Part

Assembly: Typewriting's First Impressions

"Close to a hundred inventors were doomed to build writing machines before one of them 'caught on' . . . Almost all these machines worked badly, in varying degrees of badness."

– Michael H. Adler, *The Writing Machine*

Chapter 4

Writing the Truth

The problem is where to begin.

The first sentence of the first chapter of Wilfred A. Beeching's *Century of the Typewriter*, one of the better-known popular histories of the machine, reads, "It would be impossible to write the history of the typewriter from its actual inception, for there was no true beginning."[1] This situation is not unique to typewriting. Most discourses come into being gradually, replete with many false starts, redundancies, and parallel moments of innovation as similar ideas occur to inventors and thinkers at different times and in different places. As a result, moments of origin are inevitably shrouded in myth, hyperbole, and inaccuracies of various kinds.[2] We'll have to be satisfied with approximate beginnings.

Three distinct fields of invention contribute to the appearance of typewriting: printing with movable type; the construction of automata, the early mechanical precursors to modern robots; and attempts to produce prosthetic writing devices for the disabled, especially the blind and the deaf. Each of these fields has its own logics and trajectories beside and beyond the creation of typewriting, which was in many respects a side effect of their cross-pollination. This may seem obvious in hindsight, but there is a tendency in popular histories of technology to assume that invention is a rational and orderly process with a clear goal. Examining the archive of typewriting reveals a history that is anything but orderly and systematic. It's common for histories of the typewriter to observe that the typewriter was invented at least fifty-two times; Adler argues convincingly that as many as 112 inventors may have beaten Christopher Latham Sholes, the "father" of typewriting, to the proverbial punch.[3] Even forming the sentence in this way suggests that each of the inventors producing one of those writing machines was working toward the same inevitable end, and that "the" typewriter was an identifiable object before its creation.

My approach is different. My focus is on type*writing*, which interests me more than compiling yet another history of the typewriter as a machine. The archive of documents produced by and about typewriters reveals different things than a history of mechanical invention. It reveals the moments when typewriting became a phenomenon in its own right rather than an extension of something else. It identifies the ways in which typewriting began to modify our behaviour, our social structures, and our very sense of ourselves. It even marks the moments when typewriting itself began to disappear and a new order to emerge.

Pantographs and Polygraphs

The most basic version of a machine that we could call a typewriter would be a lever with a key on one end and a printing block on the other. But before anything resembling such a device appeared, there was already the desire for a simple writing machine that could, with the aid of a lever connecting a writer to a writing surface, efficiently and legibly produce truth.

In 1647, William Petty was granted a patent for a machine that might, after only one hour of the operator's instruction, be "of great advantage to lawyers, scriveners, merchants, scholars, registrars, clerks, etcetera; it saving the labour of examination, discovering or preventing falsification, and performing the business of writing – as with ease and speed – so with privacy."[4] As a secret agent in the service of Charles I, Petty was in the business of extracting truth, and in need of a method of reproducing what he had learned in a form that could be circulated quickly, discreetly, and accurately to those who needed that information in a timely fashion.

Petty's device was actually a type of pantograph, an instrument that employs a series of arms linking two pens in order to duplicate a document while it is being written, or to trace an existing drawing. The pantographs artists employ to this day are adjustable, so that the document being copied can be reproduced at a different scale, but given the emphasis Petty placed on verisimilitude, his pantograph was probably fixed to reproduce the hand of the writer as closely as possible. Petty wasn't the first person to produce a pantograph; seventeen years earlier, Christoph Scheiner produced a similar machine.[5] The early history of writing machines is full of such devices, some of them quite elaborate. In 1762, Count Leopold von Niepperburg of Vienna built a pantograph that allowed the user to produce two or three written documents simultaneously by means of a series of connected pens.[6]

What makes Petty's pantograph worthy of note is the particular relation to the discourse of truth that he claims for it. The proper name for a fixed-width pantograph is *polygraph*. In 1803, the English-born John Isaac Hawkins received an American patent for a polygraph, which he described as an "Improvement in the pentagraph and parallel ruler."[7] When Hawkins returned to England, he turned over the rights to his invention to artist and museum director Charles Willson Peale, a friend of Thomas Jefferson. Because of that friendship, the polygraph became a major tool of one of the earliest American writers to actually produce the discourse of power and truth that still holds today. From 1804 onward, Jefferson used polygraphs for all of his correspondence. Jefferson was a classic early-adopting geek; he had two polygraphs, and continually traded in his machines for Peale's latest model (to view one of Jefferson's polygraphs in operation, see the interactive movie at the official Monticello website).[8]

The word polygraph, of course, is also the proper name for the "lie detector" law enforcement officials use in their attempts to discern whether a subject's statements are true. Electrodes connect the subject to sensors that monitor physiological changes as the subject answers a series of questions. Each sensor is connected in turn to a long mechanical arm with a pen on the end that tracks changes in those physiological processes over time on a scrolling piece of paper. Though the technology is slightly more elaborate than that of the earlier polygraph, the basic assemblage remains the same: the simplest of machines – a lever – connects a subject to a writing surface. The intervention of the machine is somehow supposed to translate the subject's inner truth into a legible written form.

Mechanical writing machines became much more sophisticated over the three centuries after Petty's invention, but the documents that described those machines invariably regarded them in the same way: levers that write the truth.

McLuhan's Whim

In *Understanding Media: The Extensions of Man*, Marshall McLuhan calls his chapter on the typewriter "Into the Age of the Iron Whim."[9] Initially, this puzzled me, because I was only familiar with the common definitions of "whim": "a fanciful or fantastic creation; a whimsical object"; "a capricious notion or fancy." I already knew that the typewriter was, in many respects, a series of capricious and even ill-advised technological decisions that had been literally cast in iron to produce the machine that is usually and maddeningly celebrated as the apotheosis of rational thinking and writing. At the same time, I wondered if that was enough to satisfy McLuhan's Joycean delight in complex puns (after all, the very first meaning the *Oxford English Dictionary* provides for "whim" is "a pun or play on words; a double meaning"). When I checked the etymological section of the *OED*, I found the following: "A machine . . . consisting of a vertical shaft containing a large drum with one or more radiating arms or beams."[10] A central mechanism with a number of linked arms describes several kinds of early typing technologies. The London Science Museum's booklet *The History and Development of Typewriters* outlines an entire genus of "Type-Wheel Design" typewriters.[11] In these sorts of writing machines, a circular plate revolves to place a given letter slug beneath a lever, which forces it down onto the paper below.[12] The central mechanisms of many of these machines could easily be described as whim-like.

Though slow in operation, this arrangement is durable and effective. It's also the basic operating principle of the devices that used to be sold for a few dollars in the classified pages of comic books, right under the X-Ray Spex. Advertising these machines as typewriters, while technically true, has the same bait-and-switch quality as the pictures of sea monkeys wearing tiny clothes and sitting around a little table.

While these type-wheel–design machines were undoubtedly antiquated by the mid-twentieth century, they are nevertheless based on concepts that made important contributions to typewriting when they first appeared. The "Mechanical Potenografo," for example, the invention of an Italian artist living in England named Celestino Galli, had two concentric rings of keys, which pushed levers down toward a drum of paper below. In an article about the Potenografo published in 1831, *The Times* noted many themes familiar from the descriptions of other writing machines: the relation to music ("It is played upon the fingers, like a musical instrument"); the machine as a means of producing written truth ("The judge on the bench may, by its means, take down the depositions of witnesses while his mind is intent upon the hearing of the evidence"); the machine as a prosthetic for the blind ("By a little habit even the blind may be made to use [it]"); speed and efficiency ("Many copies of a discourse, legibly written, may be taken at the same time"; "An instrument which will enable them to copy faster than any shorthand writer"). After this brief flash of renown, Galli and his machine – which may never have actually been built, as no one has ever claimed to have actually seen or used it – disappeared, replaced by other innovators working along the same lines (or within the same circles of influence).[13]

One of these innovators was a printer from Marseilles named Xavier Progin (or possibly, Porgin, Pogrin, Projean, or Progrin – typos existed long before typing machines), who patented his "Plume Ktypographique" (sic) in 1833. It too featured a whim-like ring design, but the operator worked the keys by pulling *up* on a series of hooks (identifiable only by a diagram on an accompanying piece of paper) that triggered the type bars below them to make an impression on the paper underneath. Progin's machine was slow to operate; it wrote, according to his own description, "almost as fast as a pen."[14] However, the Plume Ktypographique did something that none of its predecessors did: it used leverage to trigger individual type bars that converged down on a common central printing area. Moreover, this was also the first writing machine to feature proportional spacing, as its capital letters were twice as wide as the lower case. It was even possible for the operator of the Ktypographique to read what he was writing by looking down through a hole in the centre of the index to the paper below. Progin made one other contribution to the typewriter, again connecting the machine to music notation and musical instruments: he later built a machine for typing music, which utilized the key principles that led to the development of the shift key.[15]

By this point, Americans had also become obsessed with the idea of whim-like writing machines. In Massachusetts, Charles

Thurber was working on a writing machine that he called the "Patent Printer."[16] It featured forty-five plunger-style keys mounted on a circular frame. The operator spun the wheel to ink the keys, then selected the desired letter, rotated it over the platen, and depressed the key.[17] Thurber, like many of the other inventors, was intent on producing a machine for the blind.[18] The April 30, 1887, issue of *Scientific American* refers to it as "the first American type writer."

Like the other machines of its kind, Thurber's machine was massive and slow to operate (or, as *Scientific American* politely put it, "perfectly efficient except as to the element of time"). But Thurber's machine featured an important innovation: the first modern typewriter carriage, consisting of a cylindrical platen with a rack and pinion escapement. Despite this considerable advancement, Thurber's later machines moved in an entirely different, even Rube Goldbergesque direction, back toward the pantograph. In 1846, he debuted a large, loom-like machine that used "an

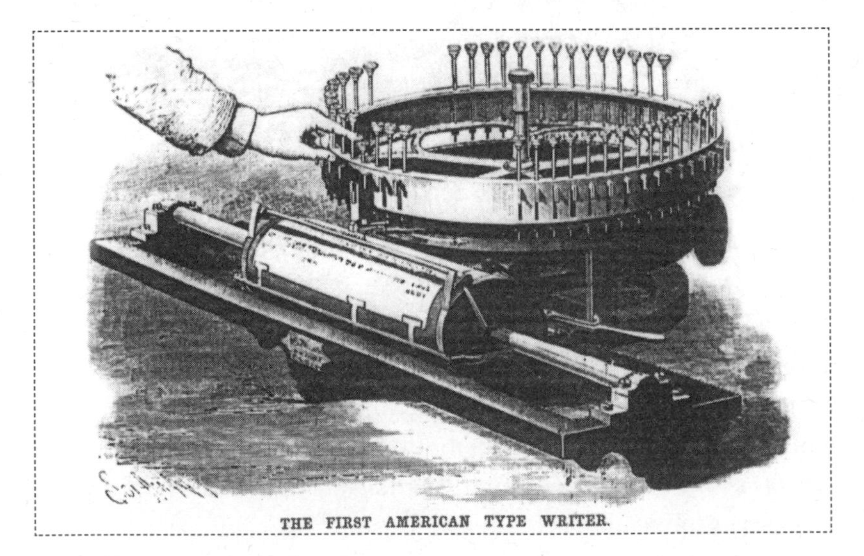

Charles Thurber's 1843 "Patent Printer," which Scientific American *later dubbed "The First American Type Writer."*

appalling collection of cams, two for each character" to write with a pencil on a vertical sheet of paper.[19] Exit Thurber.

Despite this later failure, Thurber's earlier designs had some influence. In 1852, a New Yorker named John Jones received patent no. 8980 for a "Mechanical Typographer" similar to Thurber's machine in that the letters were arranged in a circle over a cylindrical platen, but it only had one key. Nevertheless, it "incorporated the first practical use of vertically set characters" and was the first mass-produced writing machine (130 had been produced before the factory burned down – as many as the first few years of Remington's mass-produced typewriters).[20]

By 1856, the idea of a circular-indexed writing machine that stamped characters onto paper was widespread. Patent no. 14,907 went to John H. Cooper of Philadelphia for such a machine, though its lever had to be hit forcefully rather than simply pressed. Like Thurber's machine, its chief innovation was its cylindrical platen and feed roller.[21] Evidently, the editors of *Scientific American* had been paying close attention to Thurber's circular designs as well. Also in 1856, Alfred E. Beach, editor and part owner-operator of the magazine, invented and patented several writing machines featuring circular arrangements of type, including a machine for the blind. Beach's machines boasted a feature absent in Thurber's, Jones's, and Cooper's inventions: a full, three-row keyboard.[22]

McLuhan's description of the typewriter as an "iron whim," then, is a particularly satisfying combination of literal and metaphoric definitions. A whim (a capricious notion or fancy) could inspire a person to press a few keys on their whim (a typewriter as a fanciful or fantastic object), and the inspiration would be conveyed to paper by . . . a whim (a machine consisting of a central drum mechanism with a number of linked radial arms). Writing about that machine itself would be the most whimsical act of all, but it would give substance to the initial whimsies about that machine, as if those whims had been cast in iron.

An Impression upon Words

As the typewriter emerged as a machine, the poetic imagination was already preparing itself not only to write with it, but to be written on in turn.[23]

From 1756 to 1763, the British poet Christopher Smart was imprisoned in a variety of insane asylums. During that period, he wrote his famous poem *Jubilate Agno*, a linked series of fragments concerning his theories on how to assert a stable identity in what he saw as an increasingly unstable society. His description of the mechanism of inspiration is particularly interesting:

> For all the inventions of man, which are good, are the communications of Almighty God.
> For all the stars have satellites, which are terms under their respective words.
> For tiger is a word and his satellites are Griffin, Storgis, Cat and others.
> For my talent is to give an Impression upon words by punching, that when the reader casts his eye upon 'em, he takes up the image from the mould which I have made.[24]

Smart's description of the process of conveying his literary creations as "punching" words onto his readers' imaginations invokes the image of a metal slug of type punching a character onto a blank page. The extended metaphor of stars and their satellites indicates that all terrestrial objects are serial versions of a Platonic model: for example, cats are satellites of the tiger. Smart also extends this metaphor to "terms" and "words," where the word is the equivalent of the star and the terms are its satellites. The full implication of this metaphor is that Smart, too, is a satellite of God, who "punches" inspiration onto the blank page of Smart's imagination.

Smart's model of inspiration prefigures the way writers will describe writing with machines for the next two centuries. In typewriting, inspiration from "outside" (i.e., a real or imagined dictating voice) flows to the writer through the machine. Almost 250 years later, Friedrich Kittler, a major contemporary theorist of technology, writes, in a moment of philosophical rhapsody:

> I am thus a letter on the typewriter of history. I am a letter that writes itself. Strictly speaking, however, I write not that I write myself but only the letter that I am. But in writing, the world spirit apprehends itself through me, so that I, in turn, by apprehending myself, simultaneously apprehend the world spirit. I apprehend both it and myself not in thinking fashion, but – as the deed precedes the thought – in the act of writing.[25]

The mode of being that Kittler describes, that of a person obsessed with documenting every lived second of his life, is not so different from Smart's. It may even be the *cogito* of modernity: I am typewriting (both process and product); therefore I am.

Though the use of the words "punch" and "mould" connote lead type, some of the writing machines of Smart's era also functioned by a series of rods that "punched" the paper. In 1714, a British engineer named Henry Mill received a patent for a device that many sources identify as the inaugural object in typewriter history,[26] "AN ARTIFICIAL MACHINE OR METHOD FOR THE IMPRESSING OR TRANSCRIBING OF LETTERS, SINGLY OR PROGRESSIVELY ONE AFTER ANOTHER, AS IN WRITING, WHEREBY ALL WRITINGS WHATSOEVER MAY BE ENGROSSED ON PAPER OR PARCHMENT SO NEAT AND EXACT AS NOT TO BE DISTINGUISHED FROM PRINT . . . THE IMPRESSION BEING DEEPER AND MORE LASTING THAN ANY OTHER WRITING, AND NOT TO BE ERASED OR COUNTERFEITED WITHOUT MANIFEST DISCOVERY."[27] There are no known diagrams of Mill's machine, and no real evidence that he actually built it, so while its design and the process of its operation remain uncertain, what stands out in the description of it is that the force of typographic impression is associated with a non-counterfeitable, durable writing. Like Smart's divine Impressions, Mill's machine writes the truth.

The first typing machine to feature both upper and lower case characters, William Austin Burt's "Typographer" was a punching-style writing machine that developed from movable type technologies. This machine, the object of the first United States patent for a writing machine (no. 259, issued July 23, 1829), was the size and shape of a pinball machine. Its type was mounted on a semi-circular frame that was moved into place over the paper by manipulating a wheel.[28]

Burt was an accomplished inventor who made his own tools, but his first love was navigation. After inventing the Typographer, he created the solar compass (which was the standard navigational device for U.S. surveyors for seventy-five years and is the immediate ancestor of the devices currently in use) and the equatorial sextant.[29] While the Typographer was a success on its own terms, it was slow to use because it was missing a crucial element: an interface that would allow for the rapid entry of characters. In other words, a keyboard.

Chapter 5

Writing Blind

The "impressed" type of Mill's hypothetical machine had practical implications as well as philosophical ones. Raised or depressed letters have a tactile quality, and notes found in Mill's papers after his death in 1771 suggest that he may have thought of his writing machine as a device for producing texts readable by the visually impaired.[1] Several of the immediate descendants of Mill's machine, including Jean-Claude Pingeron's invention of 1780 and Franzose L'Hermina's 1784 "writing frame" (probably some relative of the polygraph), were definitely conceived as machines for the use of the blind.[2]

The first writing machine that printed in any way like a modern typewriter also stemmed from the idea of a writing machine for the blind. The laurels for this achievement go to Pellegrino Turri, who built his machine in 1808 as a favour for his patron, the Countess Carolina Fantonio da Fivizzono, who, despite losing her sight as a child, conducted a voluminous correspondence. Like many early typewriters, Turri's machine wrote only in capital letters, three millimetres in height,[3] which Count Emilio Budan describes in his 1912 essay *Precursors of the Modern Typewriter* as "thin, distinct and easy to read to this day."[4] To provide ink for the machine, Turri invented carbon paper too. Simultaneously and independently, an Englishman named Ralph Wedgwood *also* invented carbon paper – later to become an essential part of the writing assemblage as the major form of pre-Xerox duplication – and, like Turri's paper, a byproduct of his patented writing machine for the blind.[5]

Pietro Conti, a contemporary of Turri's, produced a machine in 1823 designed to allow "even those with poor sight" to write clearly and rapidly. Conti took his machine, which he called the "Tachigrafo," to France, to demonstrate and patent it. Conti's machine was solid and functional, but its various mechanical innovations were all eventually replaced by other, more effective processes. What makes Conti's machine historically important is that it was the first writing machine to be widely displayed and discussed; it inspired many subsequent inventions. His native Italy unveiled a plaque in 1934 proclaiming him "inventor of the first typewriter,"[6] but as we've seen, the lineup to claim that title is already a long one.

Like the pantographs and polygraphs, writing machines for the blind often originated in an inventor's attempt to build a machine that could read and write truth. Friedrich Kittler mentions a doctor named C. L. Müller who authored a text titled *Newly Invented Writing-Machine, with Which Everybody Can Write, Without Light, in Every Language, and regardless of One's Handwriting, Generate Essays and Bills; the Blind, Too, Can, Unlike with Previous Writing Tablets, Write Not Only with Greater Ease but Even Read Their Own Writing Afterward*. The title alone suggests that the good doctor had a rather powerful drive toward precision, but there is more. Kittler notes that "the invention was aimed primarily at educated but unfortunately blind fathers for the purpose of illuminating their morally blind sons with letters and epistolary truths."[7] In effect, the device is a prosthesis designed to convey truth, in the form of the rules of the dominant social order, between two different sorts of blind subjects.

As ideas about the potential benefits of writing machines began to spread, an increasing number of inventors began pursuing parallel ideas in different countries. In 1851, Pierre Foucault (sometimes spelled Foucauld or Foucaux) of France and an Englishman named G.A. Hughes were both awarded a medal at the Great Exhibition in London for their writing machines. Hughes picked up a second medal for his design in 1862,[8] which tied him up with Foucault, who had received a gold medal from the Paris "Board of Encouragement" before his triumph at the Great Exhibition.[9]

Foucault, one of several blind inventors who produced writing machines, built a number of them, the first of which he dubbed the "Rapigraphe" in 1839. It used a series of ten vertical punches (which combined to make all the characters in the alphabet) and made perforations in the paper rather than impressions on it. Later versions replaced the perforating points with pencil leads. By 1849, he had made a number of improvements, including the addition of sixty keys in two rows and the construction of a paper-holding frame that moved horizontally with each keystroke and could be vertically adjusted by hand to begin a new line.[10] This machine, a large, loom-like device called the "Clavier Imprimeur," was what won him the unanimously awarded gold medal at the Great Exhibition.[11]

Hughes, the director of the Henshaw Institute for the Blind in Manchester, built a small "Typograph," which was a standard tool in schools for the blind throughout the 1850s. The early model produced embossed characters, and (as was the case with Foucault's Clavier Imprimeur) later versions could also print standard letters. On a visit to Hughes's institute, Queen Victoria "was filled with astonishment and admiration" when a student named Mary Pearson typed the phrase "Her Most Gracious Majesty."[12]

Behind the invention of all of these writing machines for the blind was the unexpressed agreement that it was necessary for the blind to be able to produce text that in many cases would not even be readable by them. Nevertheless, text produced by the blind, text that supposedly told their truths, would be legible (and therefore useful) to society at large.

Part of industrialization's project was to integrate people of all sizes, shapes, and abilities as useful elements in factories or other assemblages that produced, among other things, words. As Mark Seltzer mentions in his discussion of typewriters in *Bodies and Machines*, Henry Ford's description of the perfect factory would be one staffed by the blind and amputees of various sorts, because "automatized hands work better when blind."[13] Factories are not organized around individuals; they are organized around relationships between body parts and other machines. Factories break tasks down into series of discrete steps. The equipment necessary to perform any given task may not always involve eyes, or legs, or may require only one hand. In such a setting, the hands of one worker, say, a typist, may have more in common with the hands of the people typing on either side of them than they do with the eyes of the machine's operator. The biological parts of the assemblage may eventually be replaced entirely by mechanical components, but, as Kittler notes, it is the "pitiless experiments" of nature, resulting in conditions such as blindness and deafness, that yield the information on "the human machine" necessary to imagine such new combinations of dissociated parts in the first place.[14]

The Writing Ball

Deafness also played a role in the development of the writing machine. Pastor Hans Rasmus Johann Malling Hansen,[15] head of the royal Danish Døvstummeinstitut ("Deaf and Dumb Institute") in Copenhagen, had been attempting to develop a new alphabet for the deaf, but became interested in constructing a writing machine as well. Malling Hansen conceived of his machine purely as a prosthetic designed to increase the speed of communication for those who could not speak or easily write in the conventional manner. The focus of his initial enterprise must have widened, though,

The final model of Pastor Malling Hansen's "Schreibkugel" (writing ball).

because before too long he asserted the writing machine was for use by the blind.

Hansen noted that a system dividing and allocating syllables and letters to the individual fingers could communicate much more rapidly (he estimated twelve syllables per second) than cursive writing (four syllables per second). He began to experiment with porcelain balls on which he had inscribed block letters,[16] and eventually decided on a hemispheric arrangement of first fifty-two, then fifty-four, concentric keys in what G. Tilghman Richards, author of *The History and Development of Typewriters*, refers to as a "radial-strike design."[17] In this system, the type bars act as "punches" much like those Smart describes in *Jubilate Agno*, impressing characters on a sheet of paper that scrolled below the keys on a semi-cylindrical platen.

The advantage of this arrangement was a short learning curve. Kittler cites a text claiming that "the blind, for whom this writing ball was primarily designed [could] learn writing on it in a surprisingly short time," apparently because a spherical configuration of keys was more amenable to tactile navigation than keys arranged in a plane.[18] However, it presented technical challenges for the designer, especially where clarity was an issue (and it would be, when attempting to produce characters legible by those with poor eyesight), because the keys in the outer circles of the sphere struck the paper obliquely and tended to create distorted or blurred impressions.[19]

By the winter of 1865–66, Malling Hansen had produced several "writing balls" and received a patent for them in 1870. His machines for the blind were a great success and rapidly spread to the workplace, even though he initially "did not take into account the needs of business."[20] He continued to refine his machines and briefly experimented with an electrical carriage movement in 1867, making his writing ball not only one of the first functioning typewriters, but very likely the first electric typewriter. Malling Hansen added a ribbon system in 1878, and won a gold medal at the Paris Exhibition that year, one of many awards he received for his work. According to Adler, the writing ball "may well be considered the most nearly perfect piece of precision engineering in typewriter history, past or present." Many writing balls remained in use until at least 1910.[21]

The most famous owner of one of Malling Hansen's writing balls was the philosopher Friedrich Nietzsche. The philosopher of the fragment, Nietzsche was also the first philosopher of typewriting, as Friedrich Kittler details exhaustively in *Gramophone, Film, Typewriter.*

"It is not only knowledge which has been discovered gradually and piece by piece," writes Nietzsche; "the means of knowing as such, the conditions and operations which precede knowledge in man, have been discovered gradually and piece by piece too."[22] Nietzsche's aphoristic "telegram style," argues Kittler, develops in direct response to the adoption of typewriting as a means of coping with his growing blindness and constant migraines. By 1879, Nietzsche was already considering acquiring a typewriter, and two years later, he purchased one of Pastor Hans Rasmus Johann Malling

Hansen's writing balls, which he preferred over the "too heavy" American machines[23] (it's worth noting that at the time, a writing ball weighed 75 kg, or 165 lbs.)[24]

Nietzsche knew as well that "Our writing tools are also working on our thoughts."[25] A single week of typewriting was enough to interpellate him into an assemblage with the machine that, as Henry Ford fantasized, would do away with the shortcomings of the flesh: "the eyes no longer have to do their work," wrote Nietzsche.[26]

Kittler also reproduces a poem that Nietzsche typed about his transformation into the first cyborg philosopher:

> THE WRITING BALL IS A THING LIKE ME: MADE OF
> IRON
> YET EASILY TWISTED ON JOURNEYS.
> PATIENCE AND TACT ARE REQUIRED IN ABUNDANCE,
> AS WELL AS FINE FINGERS, TO USE US.[27]

If the writing ball is a thing – an object – so too is the philosopher: "A THING LIKE ME." If the philosopher is alive, so too is the writing ball; Paul Auster was far from the first typist to suspect that this was the case. In typewriting, the organic and the inorganic fuse into "US," an assemblage of iron and meat, becoming one continuous instrument at the disposal of an entirely other set of "FINE FINGERS."

As August Dvorak, inventor of the eponymous Dvorak keyboard, put it in his book *Typewriting Behavior*, the business/vocational department of a school, where most people still learn to type, "is designed to develop you into a thinking machine capable of handling commercial tasks."[28] I'll have more to say about the way that institutions reshape bodies and machines into useful assemblages in Chapter 16. The more immediate task is to discuss how, as Nietzsche recognized with typical clarity, typewriting blurred and complicated the lines that Englightenment thinking had drawn between body and machine, inanimate and animate.

Chapter 6

Clockwork Boys and Piano Keys

While this is not a history of writing automata – and there are many fascinating books on the subject – it's worth discussing them briefly, because writers have repeatedly endowed the typewriter throughout its history with the automaton's ability to mimic life and intelligence by generating text automatically. Interestingly, accounts such as Richards's *The History and Development of Typewriters* frequently group writing automata with the various kinds of writing machines already described, as though there were no distinction between their form, intention, or functionality, though they differ dramatically in each respect.

An automaton is a machine that performs a particular task without any apparent outside control. In the sense that I'm using the term here, automata are whimsical and usually very expensive clockwork windup toys, originally produced for the amusement of the very rich, though many automata eventually toured Europe and America in displays for the amusement of the general public.

Several of the machines Richards mentions as direct antecedents to Mill's writing machine are automata, including those invented by Friedrich von Knauss, director of the Physical and Mathematical

Friedrich von Knauss's most advanced writing automaton, which could be programmed to write up to 107 consecutive words, stood nearly 8 feet tall.

Laboratory in Vienna. Between 1753 and 1760, von Knauss built four writing machines, the fourth of which was an ornate monstrosity featuring a huge wreathed base topped with four eagles. On their wings, the eagles supported a model of the solar system, which was in turn topped by the writing mechanism proper: an easel with a cherub perched on the edge and holding a pen.[1] Hidden inside the solar system model was a horizontal roller bristling with pins that activated keys corresponding to the letters of the alphabet – the mechanism that determined what the cherub wrote. When a key was pressed, the cherub dipped its quill in an inkwell and wrote the letter. After it finished writing a letter, another mechanism moved the easel to the left, and, when the cherub came to the end of a line, the easel would also shift down. The whole apparatus worked like a player piano, in that any text that it was to write had to be programmed into it in advance. Von Knauss's final model could be programmed to reproduce texts of up to 107 words in length.[2]

Swiss watchmaker Pierre Jaquet-Droz and his son Henri-Louis Jaquet-Droz also created a number of writing machines. Father and son worked together and were among the most famous artificers of automata of the eighteenth century. Among their other creations (which were lifelike enough that they landed their creators in prison as suspected magicians) they built "L'Ecrivain," a writing clockwork boy, around 1772. One of the phrases they frequently programmed into the boy was a parody of the Cartesian *cogito*: "I don't think, therefore I am not." More than two centuries after its creation, the doll still functions.[3]

Wolfgang von Kempelen, a wealthy Austro-Hungarian civil servant, amateur scientist, and courtier, invented the most infamous automaton of all time, the chess-playing "Turk," first exhibited in the court of Empress Maria-Theresa some time around 1769–70. The device, a wooden trunk four feet long and three feet high and apparently filled with clockwork machinery, took its name from the half-mannequin on its top, which wore a cloak and turban. The Turk toured an amazed Europe for most of the next thirty years, playing (and defeating) such luminaries as Napoleon Bonaparte and Benjamin Franklin. After von Kempelen's death, it eventually found its way to the Americas, and was destroyed in the Philadelphia fire of 1854. The infamy of the Turk stems from the fact that it was not an autonomous machine at all; the trunk contained room for a small man (presumably, with formidable chess skills) to hide inside and operate the mannequin as a kind of elaborate puppet.[4]

The Turk's long and fascinating history has been well documented by Tom Standage,[5] who notes that even though the automaton was, to a trained scientific eye, an obvious hoax, it was important because it acted as a catalyst for the discussion of exactly how autonomous a machine could be. While many of the authorities of the day argued that the construction of a true chess-playing machine was an impossibility, after the English mathematician and inventor Charles Babbage played two games against the device he concluded exactly the opposite, and went on to lay the foundations for modern computer science by constructing a mechanical computer he called the Difference Engine.[6]

Not all von Kempelen's creations were hoaxes. He was also known for his invention of an artificial voice box or "speaking bellows" capable of forming connected verbal utterances,[7] and for inventing several writing machines, including at least one for the blind. The primary beneficiary of this machine was a certain Fräulein von Paradies, who used it to typeset her music and writing.[8] Some sources claim that one of von Kempelen's devices was the first machine that could be called a typewriter by today's standards. The device was a box that connected keys, one for each letter of the German alphabet, to a set of hidden levers. Its spacing and punctuation were clear, and it printed cleanly. By the early years of the nineteenth century, a version of von Kempelen's writing machine had become useful enough to serve in an office setting, albeit there were only a few companies wealthy enough to afford it.[9]

Automata are actually related more closely to the Turing Machine (the conceptual machine that Alan Turing described in 1936 while attempting to precisely define what an algorithm is) and the computer than to the typewriter. What they contribute to typewriting is a furtherance of the sense of uncanniness that writing machines have had from the beginning – that some force inside or beyond the machine is actually doing the composing. The Jacquet-Droz joke "I don't think, therefore I am not" might have seemed droll in a culture where writers and philosophers sure of their sovereignty composed with a pen, but as more and more people began to write with machines, the question of exactly who – or what – was doing the thinking that produced typewriting would become more pressing.

On Another Note

In addition to advancements in pantography, automata, and machines for the blind and deaf, there is yet another parallel path in the development of writing machines. It is modelled on an arrangement of keys designed to produce sound rather than tactile or visible characters.

Many of the early writing machines featured keyboards resembling those of a piano or harpsichord. This is even true of the Sholes, Glidden, and Soulé Experimental Model from 1868, the direct precursor to the first modern typewriter, which had "handsome black walnut, piano-style keys lettered in white." But there were machines capable of writing music as early as 1745, when both Johann Friedrich Unger and a Mr. Cred invented instruments that supposedly printed on paper every note they played.[10]

While the keyboard is what enables a person using a writing machine to exceed the speed of handwriting, the musical keyboard is a poor analogy for the act of typing the Roman alphabet. A typist uses discrete letters, while a musician uses combinations of keys to produce chords. But this has never stopped people from making comparisons between the two machines. After typewriters began to be manufactured commercially, in the 1880s, turn-of-the-century advertising copy made much of the piano-typewriter analogy as a way of feminizing and domesticizing the typewriter's new and potentially threatening technology: "The type-writer is especially adapted to feminine fingers. They seem to be made for type-writing. The type-writing involves no hard labor, and no more skill than playing the piano."[11]

Some sources suggest that the first functioning typewriter to feature an assemblage of keys, rods, and inked ribbon was built by James Ranson in 1711. If true, this machine would predate even Mill's device, though it featured technological innovations that were far more sophisticated. According to a report in the *Yorkshire Post* dated May 2, 1952, a working model of Ranson's machine was displayed in an exhibition in London in 1907. The *Post* describes it as follows: "It has (or had) small keys like those on spinets and harpsichords, but each key was painted with a letter in a green, red and blue circle. Long steel rods surmounted by brass squares, engraved with letters, were attached to a steel frame. When the keys were pressed down they struck an inked ribbon. Movement was by means of powerful springs."[12]

In 1838, Antoine Dujardin built a writing machine for the purpose of recording speeches. It had twenty-six keys "resembling those of a piano," one for each letter of the alphabet (no punctuation or numerals), arranged in alphabetical order.[13] The typewriters designed by Sir Charles Wheatstone in 1851 and around 1855 both had piano-style keyboards (with an alphabetical arrangement of letters), and mechanisms that mimicked the operation of a piano. The type was mounted on a series of metal strips arranged in a semicircular fan, like the keys on a thumb-piano or the tines of a leaf-rake. When a key was depressed, a series of levers moved the type under the hammer, which forced the key onto the paper.[14] In America, a physician and inventor named Samuel Ward Francis was granted a patent in 1857 for a "Printing Machine" that consisted of a piano-style keyboard connected to a series of type-hammers arranged in a circle. Despite its overall clumsiness, the Francis machine contributed one innovation to future writing machines: a type guide.[15]

Another contender for the title of "first" inventor of a "functional" typewriter was Giuseppe Ravizza, who began thinking about building a writing machine around 1832, when he was corresponding with Pietro Conti. But it was almost fifteen years later that Ravizza began work on the first of his many writing machines, which he dubbed the "Cembalo Scrivano" or harpsichord-writer in 1847. As the machine's name suggests, its keys were piano-like, arranged in the order of frequency of use.[16] Ravizza continued work on his machines until his death in 1885, producing and selling

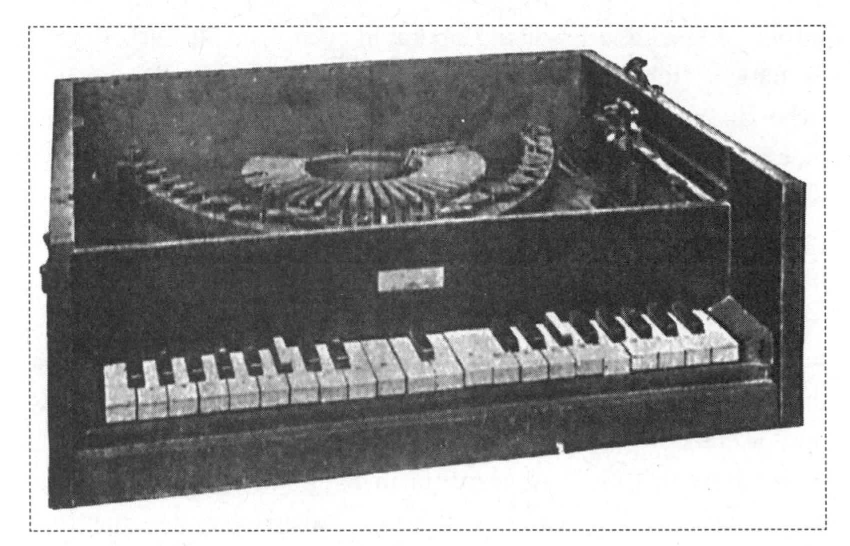

This writing machine, built by Dr. Samuel Francis in 1857, is one of the many early designs that incorporated a piano keyboard.

seventeen different models in all.[17] These machines utilized a basket-like circular arrangement of type, surrounded by a series of type bars that swung up to print on the underside of a piece of paper held in a flat frame. In 1855, Ravizza's *maccina da scivere a tasti* (writing machine with keys) appeared at the Novara Industrial Exhibition, only to suffer another first in the history of writing machines: it was destroyed by his grandchildren. Ravizza was able to rebuild this machine from the description in his patent documents, and the subsequent machine won a silver medal at the 1858 Turin Exhibition.[18]

Over the course of his career, Ravizza made a number of improvements to his various machines, including visible and audible end-of-line indicators (a little sign and a bell, respectively). The most important of these innovations was the use of a ribbon system (though Ravizza didn't actually *invent* the typewriter ribbon – that distinction belongs to Alexander Bain, who patented it in 1841), rather than ink rollers or carbon paper, to ink the keys. Like the ribbons on most of the classic modern typewriter systems, Ravizza's travelled parallel to the printing path, between two spools. He experimented with a number of inking compounds, ranging from lampblack to black lead paste, Prussian blue dye, and, eventually, a mix of aniline dye and glycerine.[19]

Close to the end of his life, Ravizza was met with a series of rude surprises and heartbreaks. E. Remington & Sons, the firearms manufacturers, had moved into the burgeoning business of making writing machines – Americans had been producing typewriters commercially for a decade by this point – and had incorporated most of Ravizza's innovations into their Model 2 typewriter. Ravizza first saw the Remington machine around 1880, only to discover that it featured many of the same innovations as his own machines (if poorly implemented; Ravizza's machine required much less effort to operate, and allowed the operator to see what he or she was typing without removing the paper). Nevertheless, Ravizza admired the operation of the Remington's cartrages and the speeds its typists achieved.[20] Ravizza continued his work, however, producing his sixteenth model in 1882, only to discover that his financial backer had lost interest. At the time of his death in 1885, he was poor and despondent, but nevertheless working on a design for a machine that wrote in syllables rather than letters.[21]

One American writing machine inventor, a lawyer named John Pratt, finished building a writing machine in 1863,during the Civil War. Pratt was a citizen of Carolina and a Confederate, and therefore ineligi-ble for a U.S. patent, so he moved to Glasgow in 1864, where he filed for a provisional British patent and began to produce prototype machines with the help of a piano manufacturing firm and a maker of scientific instruments. Predictably, the keys of Pratt's first machine resemble those of a piano keyboard; it printed by means of a hammer that pushed a carbon-paper sandwich against the type wheel. The following year, Pratt built a second model in which the type bars worked on a punch or plunger-style system, where the type bars are arranged in an inverted cone, and the type converges on a common printing point (Pratt developed this system five years before Malling Hansen).[22] Pratt subsequently completely revised his second system as well, abandoning both the plunger mechanism and the piano keyboard, which he had come to see as cumbersome, in order to develop a machine that he called the "Pterotype." In the first model of the Pterotype, the operator used six keys for each hand to control a 6×6 matrix of thirty-six characters; pressing a key with each hand moved a given character over the hammer that struck the paper against it. Pratt patented this design in 1866, but continued to make modifications. His final model had thirty-six keys and a vertical type wheel instead of a square type matrix. He exhibited his new and improved creation to the London Society of Arts in 1867. In the same year, *Scientific American* printed an illustrated editorial on the Pterotype, which described the machine as a "literary piano." According to Wilfred Beeching, this editorial also contains the first appearance of the word "typewriting." Until this point, the technology of writing machines had been described with a whole menagerie of terms: mechanical writing, typography, print writing, embossing, tachography, cryptography, and so on.[23]

From the appearance of the *Scientific American* editorial onward, society's knowledge-producers began to manufacture a discourse around the machine, its operations, and its effects. Writing about typewriting began to form something resembling a discrete body of literature, more or less separate from the material on the other kinds of writing machines that brought it into being. As typewriting grew in complexity and power, this system of rules, correlations, ordering processes, functions, and transformations[24] began to determine what typewriting can (and can't) say.

Chapter 7

The Last of the "Firsts"

By the 1860s, the climate for inventors and innovators interested in writing machines had changed considerably in two respects. People all over the world had been working on writing machines of various types for at least 150 years. Documents detailing general principles derived from that century and a half of experiments had circulated widely, and a new class of machines – typewriters proper – began to emerge as if from whole cloth. The legal climate, however, had also changed, and the inventors of this period often found themselves faced with the frustrating situation that someone else had already patented the basic principles of "their" invention.

Between 1864 and 1868, an Austrian carpenter named Peter Mitterhofer built four machines that can only be called typewriters. The question is whether Mitterhofer was an untutored genius or a clever synthesist. A rhetorical firestorm surrounds Mitterhofer and his claim to the "invention" of the typewriter. Beeching, on the one hand, baldly states that more of Mitterhofer's ideas were incorporated into subsequent typewriters than any ideas that came before him. Adler, on the other hand, argues dismissively that Mitterhofer was not first with anything, and that the

Peter Mitterhofer's typewriter, built entirely out of wood, leather, and wire.

Austrian inventor "insinuated himself" into typewriter history "on the strength of those perverse national passions with which the history of the typewriter abounds.[1] In an autobiographical poem, Mitterhofer himself claimed that "I can say with pride that the apparatus is my invention and I, a simple Tyrolean peasant from Partschins, have never seen anything that could serve as a basis for my model."[2] He never attempted to secure a manufacturer for his invention, and the significance of his work was only popularized retrospectively, in 1924, by a professor in Vienna named Dr. R. Granichstaedten-Czerva.

Because of Granichstaedten-Czerva's article, a rumour began to circulate that Carlos Glidden, who produced the "first" (there's that word again) typewriter with Christopher Latham Sholes and Samuel Soulé, had seen a Mitterhofer typewriter in Vienna and subsequently copied it. Based on the sophistication of Mitterhofer's work (and despite the fact that, according to Adler, the source of the rumor was "a 'confidence' entrusted to him [Granichstaedten-Czerva] by the son of an innkeeper in Mitterhofer's home town!"), this rumour gained enough credence that it appeared in several encyclopedias, but in 1939 Glidden's daughter testified that her father had never left the United States and knew nothing of the Austrian inventor or his work.[3] Another source alleges that Sholes stole Mitterhofer's ideas during his studies at the Royal and Imperial Polytechnical Institute in Vienna.[4] Friedrich Kittler suggests there is no evidence to back this claim, but his argument has more to do with the end of classical notions of originality and authorship that typewriting heralds: "Plagiarism, or, in modern terms, the transfer of technology, is of little importance in the face of circumstances."[5]

Mitterhofer twice petitioned Emperor Franz Joseph for funds to continue his research. The second of these petitions, written in January 1870, is of particular interest. As had nearly every inventor of earlier writing machines, Mitterhofer contended that his machine produced writing that was clear and legible; that it would be of great benefit to the blind, who would be able to learn to use it in merely a few days; that, likewise, ill people and amputees could make use of it; and, finally, that the machine would be useful for officials in the employ of the Emperor who needed to produce secret documents. These themes are all familiar, but Mitterhofer also struck some new notes concerning the efficacy of his machine for use in business or military bureaucracy. Because his machine would not, he asserted, produce undue physical strain on the machine's operator, it would be of importance to those "engaged in occupations and professions requiring mental energy e.g. solicitors, diplomats, writers." In addition, because the machine was small and portable, Mitterhofer believed it would be of great use during wartime.[6]

While Mitterhofer was attempting to elicit more than the faintest interest from his Emperor, John Pratt was returning to the United States intent on patenting and popularizing his much-lionized Pterotype. Though the U.S. patent office granted Pratt patent no. 81,000, he discovered that a handful of months earlier

The first model of Pratt's Pterotype. The operator pressed one key with each hand to specify a letter or number located in a 6×6 grid, just like in a game of Battleship.

a certain Christopher Latham Sholes had patented the machine that would earn him the laurels of being the inventor of the first commercially produced typewriter.

Pratt continued to take a drubbing at the hands of the U.S. patent system. In 1880, he applied for a patent on an improved machine that utilized a type-sleeve (the cylindrical precursor of the Selectric "golf ball" type head), only to discover that Lucien Crandall had patented the type-sleeve a year earlier. Finally, a journalist-businessman-inventor named James B. Hammond, who had also been designing typewriters, though he had yet to produce a prototype, offered Pratt a cash payment and royalty to keep his inventions to himself, and Pratt accepted. In 1884, Hammond's first typewriter model appeared, featuring many of the principles that Pratt had developed.[7]

Pratt thus joined the long, long line of technical innovators whose work has largely been forgotten, despite the significance of their contributions to the development of writing technology. Whatever the historical record may say, in the eyes of posterity invention is most often a zero-sum game. The one name that persists in the archive (though ever so faintly and not without controversy) as the "inventor" of the modern typewriter is Christopher Latham Sholes.

Sholes and Company

Sooner or later, all histories of the typewriter come around to Christopher Latham Sholes. Though the name may be obscure to most people, Sholes, at various times an editor, postmaster, state senator, and abolitionist, achieved the greatest degree of renown available to most inventors and local heroes: he has a middle school named after him in Milwaukee,[8] the city where he died in February 1890.

Sholes is generally credited with the invention of the first successfully commercially produced typewriter (who knows what would have happened if the factory producing the Jones machine hadn't burned down?). The qualifiers on that claim are crucial; as Adler notes, "there is much sick information to be examined."[9] By the time Sholes became interested in the project of creating a writing machine, the idea had been around for over 150 years, and Sholes's machine owed much to its precursors. Ravizza (1837), Beach (1855), Francis (1857), Mitterhofer (1866), and Royal E. House (1865) had all produced machines that featured a system of up-striking type bars arranged in a circular basket; Progin (1833) had used a similar type-bar arrangement with down-striking keys as early as 1833. Wheatstone (1856), Beach (in his 1856 model), and House's machines all featured keyboards with the keys arranged in rows. The machines produced by Thurber (1843), Jones (1852), Cooper (1856), Wheatstone, Francis, and House all had cylindrical platens. Many of these machines had also incorporated systems for letter-spacing and line-spacing.[10]

In 1866, Sholes and a frequent collaborator, Carlos Glidden, were attempting to build a machine that would consecutively number book folios. The design for the folio-numbering machine was based on a contraption that Sholes and another collaborator, a printer named Samuel Soulé, had built that automatically and consecutively numbered railway tickets. A fourth inventor, Henry W. Roby, was also involved to some extent. Beyond the names of the principals, though, the story becomes hopelessly muddled. Adler, who has conducted the most thorough comparison of all the existing narratives, observes that most sources conclude that after reading the *Scientific American* article on Pratt's Pterotype, Glidden showed the article to Sholes, and that the project to build a numbering machine became a project to build a writing machine as a result. Others suggest that Sholes had already conceived of the idea of a writing machine by the time that he read the Pterotype article, though they disagree about whether he had been studying other attempts to build a writing machine.[11]

One source suggests that Sholes's first mental image of the mechanism for this machine was, like that of many of his predecessors, informed by the operation of a piano. When a given key was pressed, a short type bar would swing up "in approximately the manner of a pianoforte's hammer striking from below against the strings" to hit a piece of paper.[12] However, the first model that the group produced consisted of a single type bar connected to a telegraph key. When the key was pressed, the type bar swung up to hit a carbon-paper sandwich supported by a thin glass disc. The letter on this first working typewriter key was a capital W.[13] As in "Whim."

There are two alternate versions of this story as well. One story, championed by Glidden's daughter Jennie, is that the idea originated with *Glidden*, not Sholes. In this version, corroborated by a lawyer, Glidden showed Sholes a wood-and-string model that preceded the telegraph key model. The other is that General W. D. Le Due, an enthusiast interested in the idea of a writing machine since at least 1850 and the man responsible for incorporating typewriters into the bureaucratic culture of Washington, D.C., suggested to Glidden and Sholes that they attempt to build one.[14] Whatever the impetus, in 1867, Sholes and Glidden began work on the typewriter in C. F. Kleinsteuber's machine shop, a nexus for local inventors. By fall, they had a crude prototype retrofitted from a kitchen table, and in September Sholes staged a public demonstration, which consisted of typing "C. LATHAM SHOLES, SEPT. 1867."

Bruce Bliven's *The Wonderful Writing Machine* observes that Sholes's sentence was written "in solid caps for the good and sufficient reason that the machine had no lower-case characters"[15] (and would not until the following year[16]). Sholes's choice of his own name as a first typed sentence is somewhat odd, given the claims that many popular histories of the typewriter make for Sholes's extreme modesty. *The Story of the Typewriter*, a book produced by the Herkimer County Historical Society to celebrate the fiftieth birthday of Sholes's machine, strongly suggests that self-abnegation was endemic to Sholes's personality: "As an editor he made it a rule to copy into his own paper all the adverse criticisms that were passed on him by his political adversaries, and some of them were very bitter and unjust, and he would always omit all complimentary notice of himself and his work."[17] Beeching adds that "people who knew him said that he always looked terrible whether he had money or not. He wore battered hats, a long shabby coat, and off-colour vests. His trousers were inches too short and he wore thick wool socks in low-cut shoes."[18] Richards claims, "Sholes himself was of too modest and retiring a disposition to have persevered in the face of . . . difficulties." Nor did he persevere; by the late 1880s, Sholes had totally disowned the typewriter, refusing to "own it, or even to use it or recommend it."[19] More than a little of this kind of self-abnegation remains embedded in the disciplinary regime that typewriting enforces.

The machine itself was still nameless. The major contenders were supposedly the prosaic "Printing Machine," "Writing Machine," and "Type-writer"; traditionally, this last suggestion has been attributed to Sholes. However, Adler reproduces a previously unpublished document that appears to resolve the matter: a ledger written by newspaperman and printer James Densmore in 1872. It reads, in part, "I have named the invention the 'Type Writer' and the organization or company owning it, I (think?) better be called the 'Type Writer Company'; and accordingly, I have placed that name at the head of this statement, and shall herein keep the accounts of the enterprise in that name." The invention, in other words, was christened by its financier.[20] In any event, one has to wonder about the influence of the *Scientific American* article, where the word "typewriting" first appeared.

Sholes, Glidden, and Soulé had filed a patent application for the telegraph-key prototype on October 11, 1867. For bureacratic reasons, though, the patent on this machine was not granted until a *month* after the patent for their second model, which has resulted in much confusion over the order in which the machines were produced. On May 1, 1868, Densmore filed a U.S. patent application for the second model, which used piano keys as a keyboard. "It is pointless to deny the similarities between this and later Sholes models, and some earlier machines such as Ravizza's and Francis's," writes Adler; "Whether one wishes to remain naïve enough to consider the resemblance coincidental is a matter of individual choice." The inventors improved their original concept, and by 1872, they had constructed between twenty-five and thirty separate prototypes, which they stress-tested with the help of several professional writers.[21]

The early stages of development evidently strained the financial resources of the men, because in March 1868 Densmore joined the enterprise as a financial backer. In exchange for 25 per cent of the patent shares, Densmore agreed to provide future financing and to pay all existing debts, which amounted to about six hundred dollars (the sum total of money that Densmore had at his disposal at the time[22]; he was reduced to dining secretly on raw apples and biscuits in his hotel room[23]). Critical opinion on Densmore's role and personality is sharply divided. Richards writes that "Densmore was a first-class business man of great enthusiasm, and his somewhat drastic criticism and drive was invaluable" in keeping Sholes and company on track.[24] Beeching, on the other hand, writes that Sholes had met Densmore twenty-three years earlier and taken an "instant dislike" to him, and that Densmore "made a great deal of noise and threw his considerable weight about at every possible opportunity. He was full of compulsive energy, and drove his men almost to the point of exhaustion and despair." Beeching concedes almost grudgingly that Densmore supplied "tremendous optimism" and helped to keep the project going when the others were ready to give up.[25]

While Densmore may have been an optimist, he was not an altruist. He was intent on making his fortune on the typewriter. Like Mitterhofer, he could see the machine had government and commercial applications,[26] and in 1870, he approached the Automatic Telegraph Company of New York, attempting to secure either $50,000 or $100,000 (sources vary) in exchange for the exclusive rights for its manufacture. After an initial show of interest, the telegraph company declined on the grounds that an enterprising employee named Thomas Edison was sure that he could build a better machine for less money.[27] Edison took out a patent the following year for an electric type-wheel printer, but it was closer in its design and operation to what would become the stock ticker than it was to the typewriter. It's also likely that Edison made some suggestions to Sholes for design improvements that were incorporated into subsequent models, including a 90° rotation of the carriage to the position found on modern typewriters.[28]

Densmore had other cards up his sleeve. Since 1870, he had been wooing the attentions of George Washington Newton Yöst, a petroleum salesman, agricultural implement inventor, and factory owner. Yöst was intrigued enough that he sold his factory to raise funds to invest in the machine. Though the exact circumstances precipitating Yöst's involvement remain unclear, his effect on the enterprise was immediate and galvanizing. Densmore may have been optimistic and energetic, but was likely a difficult human being. Yöst, on the other hand, was smooth, charismatic, and could sell anything to anyone – even the typewriter.[29]

It was the team of Densmore and Yöst that finally secured a sale of the typewriter in March 1873 to a major company, E. Remington & Sons. Though Remington began as a manufacturer of small arms, by 1873 they were already well established in the manufacture of other items, especially sewing machines and farm equipment. Beeching comments with a lame attempt at wry humour that "now that the Civil War was over and most of the Indians had been shot, there was very little demand for guns!"[30] Bliven's language and sentiments are similar, if more politic, suggesting that Remington was looking to diversify business because "after the Civil War boom things had been on the slow side."[31] And yet, as Mitterhofer predicted, the typewriter would eventually play an important role in both the ongoing history of manufacturing itself and in the military arsenal.

It was H. H. Benedict, the treasurer of the company's sewing machine department, and W. J. Jenne, the works superintendent, who convinced company president Philo Remington of the typewriter's potential value, and on March 1, 1873, the company signed a contract for manufacture with Densmore and Yöst.[32] Remington dedicated a full wing of their manufacturing plant to the production of typewriters, assigning Jenne other mechanics, including Jefferson M. Clough and B. A. Brooks,[33] to the project.[34] Yöst subsequently joined Remington as part of this team, and stayed with the company until 1878.[35] The company began to produce typewriters in September 1873, and shipped the first commercial machines, known as the Sholes and Glidden typewriter, in 1874. By 1876, it had been renamed the No. 1 Remington, and two years later, when the case design was changed, the No. 2 Remington.[36]

By the time Remington was mass-producing typewriters, another factor had entered the discourse of typewriting: the question of who would actually be operating the machines. Even though Remington was still first and foremost an arms manufacturer, it wasn't the military applications of the typewriter that initially determined the aesthetics of production. After all, it was the treasurer of the sewing machine department who was leading the typewriter project. Shortly after the introduction of the No. 1 Remington, the machine was manufactured on an accompanying sewing machine stand, with the treadle operating the carriage return. The colourful flowers that adorned the black lacquer Remington sewing machines also found their way onto the Remington typewriter. "The rationale behind the decorations," writes Thomas A. Russo in *Mechanical Typewriters: Their History, Value, and Legacy*, "was that the women would be less intimidated by the fact that it was a machine."[37]

Both the treadle and the enamelled flowers proved to be temporary embellishments. What no one expected was that the relationship between women and typewriting would be so successful that it would change not only the face of the workforce, but the manner in which authorship, and ultimately even authority itself, was structured.

Part

Amanuensis: Typewriting and Dictation

"I've been banging on this bunch of keys for so long they know what I'm going to write before I do."

– Barbara Gowdy, *The Romantic*

Chapter 8

Taking Dictation

Typewriting always begins with something telling someone what to write.

"Follow these instructions as though they were being dictated," commands the first sentence of Ruth Ben'Ary's famous textbook *Touch Typing in Ten Lessons*,[1] written in 1945 but still in common use today. Ben'Ary's command spells out one of the secret rules of typewriting: at some point, even a lone "generative" typist has to learn to type by following someone else's dictation, without question. Sometimes a book or an instructional audiotape or a piece of software or even a half-assed personal notion of how to hunt-and-peck one's way across a keyboard substitute for the stern voice of the high-school typing teacher. All of these possibilities amount to the same thing: someone or something, even if it's just another part of ourselves, dictates to us, tells us what to write until we internalize it and forget about it. Even then, the dictatorial voice that makes typewriting possible very often comes back to haunt the typist, after being split, stretched, twisted, and transmogrified into something uncanny and alien by the typist's imagination. In other words, no one is ever alone at a typewriter.

What actually produces typewriting turns out to be a surprisingly variable assemblage of people and machines. From the relative beginnings of typewriting, this assemblage has consisted of three positions. There is a space for a *dictator* – the source of the words that are being typed. There is a space for a *typewriter* – that is, an actual writing machine. And there is a space for the person who is actually operating the machine – an *amanuensis* ("One who copies or writes from the dictation of another," from the Latin for "hand servant" + "belonging to").[2] The problem is, it can be very difficult to determine who – or what – is occupying any of those positions at any given time.

Communication always depends on two empty spaces that we take turns occupying: "I" and "you." When I speak, I'm "I" and you're "you," and when you speak, it's your turn to be "I."[3] In typewriting, "I" (the dictator) and "you" (the amanuensis) continue to oscillate back and forth. Avital Ronell, who has written extensively on the subject of dictation, uses the analogy of the tight twin spiral of a DNA molecule to describe the relationship between the dictator and the amanuensis,[4] because dictation weaves them together in a manner that makes their writing styles indistinguishable.[5]

In some cases, dictator and amanuensis can and do change positions, or a new dictator or amanuensis can take up where the previous one has left off, all without leaving any clues as to this occurrence in the typescript. The amanuensis can also change the dictator's words, deliberately or accidentally. In any event, "I" and "you" create a typewritten document together, and from reading that document, it's usually impossible to tell whose words finally ended up on the paper. Typewriting confuses you and I. In his analysis of Franz Kafka's first typewritten letter, Friedrich Kittler spots twelve typos, over a third of which involved the German equivalents of "I" or "you," leading Kittler to observe that it's "as if the typing hand could inscribe everything except the two bodies on either end of the . . . channel."[6]

What's more, "I" and "you" are never equal. I tell you what to do – "Follow these instructions as though they were being dictated" – which is not to say that you will actually type what I tell you to. In any relationship, there are always power imbalances, and there are always struggles. Typewriting is rarely simple, but it is never innocent.

Haunted Writing

From a philosophical point of view, *all* writing begins with the process of dictation. No matter what arbitrary starting point we choose – Genesis 1:3 ("And God said, Let there be light: and there was light") or Plato transcribing the lectures and dialogues of Socrates – something dictates, and someone writes that dictation down. But what happens when the dictator dies?

Dictation continues to function even (especially?) when the dictator is absent or dead. There is nothing to stop the amanuensis from signing the dictator's name on what they have just "transcribed" . . . and there never has been. In the name of assuming the considerable responsibility of conveying the thoughts and ideas of the dictator to a wider audience, the amanuensis can actually rise up and turn the dictator into the equivalent of a ventriloquist's dummy. Jacques Derrida notes that this bald fact requires us to have a certain amount of scepticism, and to retain our sense of humour, about even the most hallowed of dictated texts while continuing to take them very seriously; since, when Socrates "dictates" his work to Plato, "*Socrates* was already no longer there and moreover was never asked for his opinion, you can see what we have been working on for twenty-five centuries!"[7]

Simultaneously taking and not taking dictation seriously is the only way to make sense out of the various odd scenarios that the long history of dictated writing presents to the reader. Plato "transcribed" the words of his dead friend Socrates. Poets throughout history have claimed that the words that they write came not from their own minds but from the inspiration of various muses. The German writer Johann Peter Eckermann, the friend and assistant of Johann Wolfgang von Goethe, continued to receive dictations from Goethe long after Goethe's death. Science fiction writer Philip K. Dick wrote millions of words about the texts that were beamed directly into his brain by VALIS, a "Vast Active Living Intelligent System" built by aliens and orbiting somewhere out on the edge of the solar system . . . and so on. Regardless of whether the odd "dictators" of these works are literary conceits, the products of fevered imaginations, or deliberate deceptions, the works themselves remain valuable contributions to culture.

Typewriting makes the question of authorship even more difficult to determine because it removes even the illusory certainty that a handwritten manuscript offers. Something dictates, someone types, and a page of standardized mechanical text appears. The only thing that is always visible in typewriting is the typewriter itself. Suddenly, incontrovertibly, there is a machine between the dictator and the amanuensis . . . and it consumes both of them as each tries to consume the other.

Chapter 9

The Typewriter Between

As the outlines of the typewriter become more distinct in the cultural imagination, those of the people surrounding it begin to dim, as though the machine were draining the vitality from its human users in order to strengthen itself.

In *Laws of Media*, Marshall McLuhan writes that although the typewriter is "a merely mechanical form," it "acted in some respects as an implosion, rather than an explosion."[1] After making this observation, particularly in a book subtitled *The Extensions of Man*, it is somewhat puzzling that McLuhan doesn't explore it further. If he had, he might well have come to the conclusion that typewriters don't extend humans; humans extend typewriting.

As typewriting became an integral part of the modern workplace, it began to affect language itself. At first, the word "typewriter" referred to the amanuensis; the machine itself was always the "typewriting machine." But as popular histories of the technology never tire of observing, "Ladies who operated Type-Writers were quickly nicknamed 'typewriters.'"[2] In this sentence and forever after, the machine deserves the proper noun, and its endlessly replaceable human operator becomes a lower-case namesake.

Words persist, but their meanings change. The "Secretary" may well be the man in this image, in which case the "Typewriter" refers to the woman and *the machine, separately and as an assemblage.*

On the other side of the typewriting assemblage, the dictator sometimes achieves a degree of mastery and power over writing undreamt of in the days of the pen. William Roscoe Thayer's 1919 biography of Theodore Roosevelt details the transformation that the typewriter helped to create in the White House bureaucracy between the Cleveland presidency and the Roosevelt presidency:

> Only a few years before, under President Cleveland, a single telephone sufficed for the White House, and as the telephone operator stopped work at six o'clock, the President himself or some member of his family had to answer calls during the evening. A single secretary wrote in long hand most of the Presidential correspondence. Examples of similar primitiveness might be found almost everywhere, and the older generation seemed to imagine that a certain slipshod and dozing quality belonged to the very idea of Democracy.
>
> Nevertheless this was a time of transition, and the vigor which emanated from the young President passed like electricity through all lines and hastened the change . . . Instead of one telephone there were many working night and day, and instead of a single longhand secretary, there were a score of stenographers and typists.[3]

In this passage, typewriting is synonymous with an alert, industrious, competent, and fully technologized modern office, powered by the electric "vigor" of the young Roosevelt. And any "president" worthy of the name, implies H. L. Mencken in *The American Language* (written during the same period as the Thayer book), surely has at least two typists at his disposal.[4]

But this is only one version of typewriting, a male fantasy about the power that the machine can bestow. As the logic of typewriting takes hold, the dictator, too, disappears into the assemblage . . . as though the machine were somehow . . . vampiric.

Chapter 10

Transylvanian Typing

Inserting a typewriter into the scene of writing affects all of the rules under which dictation takes place. One of the most obvious changes is a drastic acceleration of the rate at which dictator and amanuensis possess and consume each other. One of the most compelling demonstrations of this aspect of typewriting is Bram Stoker's *Dracula*.

Dracula, argues literary scholar Jennifer Wicke, *is* typewriting, and typewriting is vampiric. Wicke suggests that *Dracula* is the first great modern novel because it tells its story through a kaleidoscopic array of what would have been, at the time of its production, the cutting-edge media technologies that were already making up the stuff of mass culture.[1] The uncanny powers of the Count mirror the emerging powers of mass media, which goes a long way toward explaining on the one hand why we are compelled to tell this story over and over again,[2] and on the other why, like the vampire's non-existent reflection, *none* of the thirty or so film versions of *Dracula* depict the novel's characters' obsessive use of typewriters, stenography, Kodak cameras, and phonographic dictation to document their every move.[3]

A "Note" appended to the novel and signed by Jonathan Harker, one of the principal characters, states the following: "We were struck with that fact, that in all the mass of material of which the record [i.e., the text of the story] is composed, there is hardly one authentic document; nothing but a mass of typewriting."[4] Just as Dracula cannot reproduce other than by draining the blood from the living, creating inferior replica vampires in the process, *Dracula* the novel is only ever a collection of multi-generational typescripts . . . copies of copies. The Count himself has seen to that. At one point late in the novel, he breaks into the protagonists' chambers and throws their typescripts, and even Dr. Seward's diary, dictated onto wax phonograph cylinders, into the fire, as if to destroy anything resembling a definitive firsthand account of his actions.[5] Unbeknownst to Dracula, though, everything has been typed "manifold" – presumably, meaning with two sheets of carbon paper and two flimsy copies – resulting in three copies of every typescript.[6] Wicke notes that the manifold function of the typewriter brings the story "into the age of mechanical reproduction with a vengeance," as it evokes Dracula's many disguises and pseudonyms, as well as the process by which he produces more vampires.[7] Typewriting may be vampiric, but it can also be used to fight vampires.

The person responsible for that mass of typewriting is Miss Mina Harker, assistant schoolteacher and fiancée of Jonathan Harker. She sees the role of wife and amanuensis as one and the same, as she explains in a letter to her friend Lucy early in the story: "When we are married I shall be able to be useful to Jonathan, and if I can stenograph well enough I can take down what he wants to say in this way and write it out for him on the typewriter, at which also I am practising very hard."[8] In the name of producing a complete record, the better with which to combat the vampire, Mina dutifully types not only her husband's shorthand diaries, Dr. Seward's phonograph journal, and the letters of their other allies, but also a wide variety of newspaper clippings containing items that might relate to Dracula's actions. Sometimes, as in the case of the phonographic cylinders, typewriting vampirically drains these sources of their specificity. After listening to Seward's journal, Mina says to him, "That [phonograph] is a wonderful machine, but it is cruelly true. It told me, in its very tones, the anguish of your heart ... No one must hear them spoken ever again! See, I have tried to be useful. I have copied out the words on my typewriter, and none other need now hear your heart beat, as I did."[9] Sometimes, as in the case of Jonathan Harker's stenographic journal, typewriting infuses an abbreviated code with an uncanny vitality, turning a series of cryptic scribbles back into a fully realized narrative stream just as a meal of blood flushes a vampire's dead cheeks.[10] For Wicke, the inclusion of extensive transcriptions of newspaper articles are the most telling of all, because they mark the novel's parasitic reliance on the emerging narratives of mass media: "Even at the narrative level *Dracula* requires an immersion into mass-cultural discourse; its singular voices, however technologically-assisted, are in themselves not sufficient to fully explicate the vampire's actions."[11] Covering the Count is necessarily a multimedia activity. In all cases, the result is an undifferentiated flow of type, a merging of a multiplicity of many dictating voices into a flow of mass-mediated content.

Among those dictating voices is Dracula himself. By both feeding on Mina Harker's blood[12] and then forcing her to drink from his own,[13] the Count creates a binding parasitical relationship between Mina and himself. The implication is that she is at least part way to becoming a vampire, and is mentally linked to Dracula. Wicke notes that, paradoxically, it is the act of being parasitized that gives Mina authority in every sense of the word.

> After her vamping, the men alternately need to tell her everything, and want to tell her nothing. Oscillating back and forth between these positions, Mina becomes more and more the

> author of the text; she takes over huge stretches of its narration, she is responsible for giving her vampire-hunting colleagues all information on Dracula's whereabouts, and she is still the one who coordinates and collates the manuscripts, although she has pledged the men to kill her if she becomes too vampiric in the course of time. Her act of collation is by no means strictly secretarial, either; Mina is the one who has the idea of looking back over the assembled manuscripts for clues to Dracula's habits and his future plans. Despite the continual attempts both consciously by the characters and unconsciously by the text itself to view Mina as a medium of transmission, it continually emerges that there is no such thing as passive transmission . . .[14]

Dracula's defeat brings with it mixed blessings for Mina. On the one hand, she is no longer in danger of becoming a vampire, and "is free to become a mother, to reproduce what she has heretofore only copied."[15] On the other hand, without her typewriting equipment, she has no voice at all, and the last word in the novel goes to her husband.

There is one final act of vampirism, though: the reader's consumption of the text. If Dracula is a figure for mass media, *Dracula* is itself the story of mass media, of how we came to be creatures of perpetual consumption, always hungry for new flows of information but never satiated. As Wicke sardonically concludes, "Under the sign of modernity we are vampires at a banquet of ourselves."[16]

Where typewriting is concerned, the question of who is being consumed and who is doing the consuming has always been particularly vexing for women.

Chapter 11

The Type-Writer Girl

In the Romantic era that preceded modernism, which Friedrich Kittler christens "the Age of Goethe" after one of its most prominent figures, writing (both professional and literary) was done with a pen, and, as the last century of literary and cultural criticism has described, writing in the Age of Goethe was a largely male enterprise. When the entire production process was dominated by men, the only position left for women, Kittler argues, was in the audience.[1] The sheer amount of text that modern business methods required in order to function, however, was about to change everything.

At the turn of the century, the emergence of large corporations and global markets produced a blizzard of documents – accounting ledgers, purchase orders, memos, correspondence, and so on – which in turn required increasing numbers of clerical workers to produce, reproduce, sort, and file these documents. The people that began to fill these roles were educated, middle-class women.

The statistics on the sex of professional stenographers and typists in the United States from 1870 to 1930 demonstrate a startling transformation of clerical labour.[2] In 1870, 4 per cent of typists were women. A decade later, in 1880 (when the Remington No. 2 first hit the market), that number had jumped to 40 per cent. This explosion didn't escape the notice of the institutions responsible for producing useful members of society for very long; in 1881, the Young Women's Christian Association began its first typing class for girls, with eight students.[3] It also became a common practice for typewriter manufacturers to establish typing training programs for young women "and then more or less 'sell' them to business houses with their machines."[4] Typewriter advertising adopted a similar strategy, hiring fashionable young women with just enough typing skills to demonstrate the product in a showroom setting. This practice led to competing firms touring their spokesmodels on lecture and exhibition circuits, a kind of precursor to the Budweiser Girl.[5] By 1910, 80.6 per cent of typists were women, and by 1930, almost *all* typists (95.6 per cent) were women.

Many books about typewriters repeat G. K. Chesterton's quip about this turn of events: "women refused to be dictated to and went out and became stenographers."[6] Chesterton's joke raises an important point about the nature of power, which is never entirely oppressive. At the same time as it shapes and controls and coerces us, power is also what creates the skills that allow for rebellion against and reform of the institutions that wield that power.[7] The typewriter did not so much produce or repress the emancipation of women as it redistributed the regimes of control throughout society. As women became part of the industrial workforce, there were losses of power for them, but there were also gains.

Different sources present the battle for women's rights in the workplace in radically different terms. The Herkimer County Historical Society's *The Story of the Typewriter*, taking a tone common to the early histories of the machine, insists to the point of incredulity that the typewriter was the major means of women's emancipation. For example, it repeatedly asserts that the typewriter "freed the world from pen slavery,"[8] and, bizarrely, that

Christopher Latham Sholes represents the best "choice of some historic figure to symbolize [the feminist] movement."[9] Richard N. Current's *The Typewriter and the Men Who Made It* is slightly less hyperbolic: "No invention has opened for women so broad and easy an avenue to profitable and suitable employment as the 'Type-Writer,' and it merits the careful consideration of all thoughtful and charitable persons interested in the subject of work for women."[10] Current's phrase "suitable employment," however, merits further discussion.

When the YWCA formulated its plan to begin teaching young women to type, the popular consensus was that typing was anything but suitable employment. In his detailed essay "The Cultural Work of the Type-Writer Girl," Christopher Keep notes that the public reacted to the Y's plans as if the apocalypse were nigh. Many people believed that the women who became typists personally risked "unsexing" themselves, and might eventually experience a complete mental and physical collapse. Others predicted even more dire consequences, up to and including the collapse of the family unit and the moral integrity of the nation.[11]

Percentages of Girls of Various Ages Who Report That They Would "Like Best" to Be (*a*) Stenographers or Typists, (*b*) Movie Actresses
(From Lehman and Witty [32])

Age in Years	Typists	Movie Actresses
10½	14%	29%
12½	26%	17%
14½	31%	10%
16½	32%	5%

Dvorak included this chart from a serious study in a business trade journal in his book Typewriting Behavior *as an indication of the importance of typing as a vocational skill. Today, it indicates how impoverished the career choices for young women have been.*

In the disapproving eyes of the late nineteenth century, the simple need for more clerical workers was not enough to legitimize women's entry into the workforce. What made it possible for society to consider the role of typist as "suitable employment" for women was the development of a positive association between women and the typewriter.[12]

Writing machine and amanuensis are represented by a single word ("typewriter") because they are a package deal: each requires the other in order to function. The merging of the two novelties (working woman and weird gadget) alleviated the suspicion that either on their own might have elicited. While the typewriter introduced a system of discipline that moulded women workers into a form amenable to the needs of the corporate environment, the typewriter also made it possible for women to overcome many of the gender-based restrictions that were a traditional part of writing.

The novels, plays, short stories, music hall routines, illustrated advertisements, and postcards of this period began to feature an entirely new creature designed to lure women into the workplace: "The Type-Writer Girl." Her debut may well have been in a series of letters that Rudyard Kipling wrote in the United States between 1887 and 1889 for publication in a pair of journals in India.[13] This may seem odd at first, but after all, Kipling was on familiar territory, reporting on sightings of new and exotic peoples and professions for the readers of the Empire. In fact, his conclusion regarding how to deal with the Type-Writer Girl, while presented in a humorous light, is a classic colonialist response.

Kipling notes that while the Type-Writer Girl was "an institution of which the comic papers make much capital," she was nevertheless "vastly convenient." What puzzles Kipling about the Type-Writer Girls is that while he suspects that there is still very little difference between American women and their English counterparts "in instinct" (namely, they are uninterested in working for a living and are merely waiting until a suitable husband approaches them), the way that they speak and act indicates that they behave according to an entirely different set of rules. When Kipling finally does locate one female typist who admits to being interested in leaving her job for a prospective husband, and is about to consider her proof of his original thesis, she responds by quoting bon mots from French literature to him, which leaves him completely bemused: "What is one to say to a young lady . . . who earns her own bread, and very naturally hates the employ, and slings out-of-the-way quotations at your head. That one falls in love with her goes without saying; but that is not enough. A mission should be established."[14]

If one of Kipling's conservative contemporaries were to form a mental image of the Type-Writer Girl based on her depiction in the literature of the time, they might well conclude that a mission was indeed necessary. In Grant Allen's 1894 novel *The Type-Writer Girl*, the eponymous protagonist embarks on a series of titillating activities, including bicycle-riding, smoking, and cavorting with anarchists in the English countryside.[15] J. M. Barrie's 1910 one-act play "The Twelve-Pound Look" (the New Woman's analogue to the Vietnam Vet's Thousand-Yard Stare) reads like a sequel to Ibsen's *A Doll's House*: wealthy businessman Harry Sims discovers that the Type-Writer Girl he has just hired from the Flora Type-Writing Agency is none other than his estranged first wife, Kate. When Harry, now worth a quarter-million pounds, tries to lord it over her, Kate replies, "I'll tell you what you are worth to me: exactly twelve pounds. For I made up my mind that I could launch myself on the world alone if I first proved my mettle by earning twelve pounds; and as soon as I had earned it I left you."[16] Harry, of course, dismisses her scornfully, but the script implies that he will receive his comeuppance once again, as, at the end of the play, his fiancée asks, "Are they very expensive . . . those machines?"[17]

Hollywood's take on the Type-Writer Girl is *The Shocking Miss Pilgrim*,[18] a musical directed by George Seaton, with lyrics by Ira Gershwin, and starring Betty Grable as the eponymous shocking typist. The plot unfolds much as one would expect: in the late 1800s, young Cynthia Pilgrim completes her training as a Type-Writer Girl and becomes the first female employee at a Boston shipping company. Despite Cynthia's suffragette politics, she is soon embroiled in a romantic relationship with her employer. Hijinks ensue.[19]

In an early instance of product-placement advertising, Remington Rand launched an extensive series of print advertisement tie-ins. Under the headline "Shocking in 1873 . . . Essential TODAY," one ad shows Grable in character and out, seated at an antique Remington covered with lacquered flowers and at a contemporary business machine by turns. The ad's text unabashedly and abruptly co-opts the rhetoric of suffrage into a sales pitch:

> *The Shocking Miss Pilgrim*'s granddaughters . . . the millions of typists of today . . . have made the American office a warmer, more human place. Vital part of the national economy . . . without the typist, the office as we know it today just couldn't exist. Like Miss Pilgrim, the modern typist knows her work is easier with the new Remington typewriter . . . it goes faster, more smoothly, and that now, with Keyboard Margin Control*, setting margins is simple – all she has to do is "Flick the Key – Set the Margin!" Modern business men, too, like the new KMC* Remington . . . for its beautiful typescript, for its operating efficiency, for its flexibility.[20]

While the female typist has to worry about actual operations, the "Modern business man" can evidently concern himself with the aesthetics of his typewriter(s) at work.

More interesting yet is the fact that *The Shocking Miss Pilgrim* started out as an altogether more serious project. As she details in her memoir (also titled *The Shocking Miss Pilgrim*[21]), screenwriter Frederica Sagor Maas's original script was titled *Miss Pilgrim's Progress*, a thought-provoking drama about the entry of women into the workplace. The dumbing-down of the story became, instead, an allegory for Maas's experience as a woman working as a writer in Hollywood; she was almost immediately labeled a troublemaker by studio executives and "had difficulty finding work, despite earlier successes."[22]

While the life of the female typist was developing a definite cachet in the popular media, the missionaries of the new secular economy ensured that it was not all sunshine and roses for the Type-Writer Girls. Newly emerging corporations structured their wage systems and working environment to ensure a maximum amount of control over single working women.

Type-Writer Girls received extremely low wages. A survey for *The Economic Journal* found that in 1906 the average weekly wage of a female typist in England was between twenty-five and thirty shillings.[23] The memoirs of Janet Courtney, one of the first female clerks at the Bank of England, recall that this amounted to a bare subsistence living, because most hostels and boarding houses of the time charged about twenty-five shillings a week for room and partial board[24] (the American Type-Writer Girls with whom Kipling spoke were apparently in the habit of living two to a room in the business district to cut costs and avoid paying for transit). Businesses made a widely accepted distinction between "individual wages" and "family wages," arguing that while single women had only themselves to support, most male workers had to also support a family. This logic, however disingenuous, suggested that it was in a working woman's best interests to earn between 25 and 50 per cent lower than her male counterparts, because to accept more money for her work would mean not only that married women and their children would suffer, but the number of men earning enough money to take the Type-Writer Girl for a wife would be diminished.

As more Type-Writer Girls joined the workplace, corporations restructured themselves to ensure that these women would rarely if ever enter the management stream.

> The once inclusive category of clerk was increasingly subdivided between those tasks which required "decision-making" skills and those, like typing, which were merely "mechanical" in nature. This distinction masked what was in reality a division of labor along gender lines: men, who were felt to possess superior intellectual abilities and greater strength of character, continued to be placed in positions which allowed them to rise in the administrative ranks, while women were confined to jobs which were in effect occupational dead ends.[25]

Both the civil service and private corporations were thus able to amass large pools of educated, talented, but nevertheless cheap labour while simultaneously being able to insist that they were looking out for the welfare of the family unit and society at large. Meanwhile, the Type-Writer Girls scratched out the meanest of livings, while their glamorous image attracted a steady stream of new women into the workforce.

Why the disparity between the reality of the life of women typists and their portrayal? Critic Leah Price notes that Victorian fiction and culture in general promoted a double standard that made professional ambition a vice for men and a virtue for women. If a novel or story presented a woman as an author, it would often assign her "an emasculated hireling"; the secretary, by contrast, always had "a manly boss" (shades of Mina Harker in *Dracula*, whose authorial typing makes her more than a little unnatural – Professor Van Helsing describes her as having a "man's brain . . . and a woman's heart"[26]). The implication is that for women the correct side of the typewriter is the side with the secretary's chair, and that they should limit themselves to anonymous passive transcription rather than presuming to actually dictate.[27] While popular imagery exaggerated the independence of the Type-Writer Girl, it did so only to imply that part of her longed to be swept away by the right man. By initially presenting her as an exotic new species requiring something akin to the efforts of a missionary to "convert" back to the orthodox roles of wife and mother, fiction and advertising alike turned the Type-Writer Girl into something of a fetish object.[28]

Beeching mentions the many "predictable Music Hall gags about men working with typewriters on their knees."[29] Bliven concurs that merely using the word "typewriter" with a leering tone of voice was enough to bring down the house,[30] and Christopher Keep observes that there was soon a thriving cottage industry of Type-Writer Girl pornography.

Predictably, there are plenty of examples of "Tijuana Bibles" – small, cheaply printed pornographic comic booklets produced in Mexico, featuring crudely drawn images of popular comic strip characters shagging their brains out – with Type-Writer Girls in the starring roles. Just as predictably, some of these booklets have been digitized for posterity and now appear on various websites. Tijuanabibles.org hosts at least three pieces of Type-Writer Girl

"Taking dictation," Tijuana-style.

porn, two based on the adventures of "Smitty the Office Boy" with the boss's stenographer.[31] The real Smitty was the creation of Walter Berndt, who drew the strip for nearly sixty years, beginning around 1920, but his characters' poorly drawn counterfeit cousins fare much better in amorous matters. The roles in all cases are predictable. The lecherous boss leads with Music Hall-style innuendo: "Miss Higsby, are you ready for – ahem! – er – dictation." The Type-Writer Girl barters sex for more money, with varying degrees of cynical acumen. In one bible, she is demure and suggestive: "Mr. Smith, I'd do anything to get my wages increased"; in another, she is hardened and sarcastic: "[Bossman:] Well? How about it, Miss Titts? Five bucks more a week if you act nice. [Steno:] Well, looking at that thing of yours I'd say ten and I'll consider it." Smitty, of course, services the steno for free, just as he did her predecessor.

The image of the wanton Type-Writer Girl evolved as an attempt to delineate the differences between men and women in the workplace: that regardless of their new roles, they were still primarily sexual objects subordinate to the desires of men.[32] As typewriting insinuated itself further into the workings of culture, those lines became harder and harder to maintain. In the world of literary authorship, where the lines had been among the firmest, they were also among the first to falter.

Chapter 12

Remington Priestesses

The image of the great masters of Western literature dictating their most famous works to female secretaries also has some of the fetishistic quality of the Type-Writer Girl mythology. While there exist plenty of photos of writers dictating to typists, it is almost impossible to be sure exactly what is being dictated, yet there is a will to present those scenes as the scenes of great writing. Beeching's *Century of the Typewriter*, for example, refers to a photograph showing "Tolstoy dictating – one of his novels(?) – to his daughter on the typewriter." The caption accompanying the photograph goes further, claiming that in the picture Tolstoy is "said to be dictating his novel *War and Peace* to his daughter."[1]

As Sandra Gilbert and Susan Gubar detail in *The Madwoman in the Attic*, there is a long tradition of women secretaries serving in the same capacity as John Milton's daughters: taking dictation for the blind poet and tending to his every material need. This is a far from idyllic occupation; Gilbert and Gubar's critique of this relationship turned "Milton's Daughters" into a metaphor for the suppression of women writers. As much as she was anything else, the Type-Writer Girl was "Milton's great-granddaughter," and

Remington Notes

Volume 2 New York Number 7

COUNT TOLSTOY AND HIS THREE ASSISTANTS

Two women + one machine = Tolstoy's three assistants.

suffered all the hardship that that epithet implies, even though sometimes the typewriter was a mechanism that finally permitted her to talk back.

The Typewriter: History and Encyclopedia, a trade publication from 1924, features a photo of Christopher Latham Sholes's daughter seated at an early typewriter. Though she was ostensibly "the first woman who ever wrote on a typewriter" and "she made a prominent figure in typewriter history through participation in the Ilion celebration [of the commercial typewriter's fiftieth birthday, September 12, 1923], which she attended in company with her husband," the account at no point mentions her given name, referring to her only as "Mrs. Charles L. Fortier."[2] What makes the lack of her given name particularly striking is that Sholes's own name was (according to legend if not in fact) the first thing he typed on his first complete typewriter, though he was supposedly an extremely self-effacing man.

Typewriting comes into being at the same time as the notion of public authorship, aiding and abetting it. Leah Price observes that the first generation of typists was trained during the same period that anonymous publication was falling out of favour, and authors (with the exception of some journalists) were beginning to sign their legal names to their work.[3] The typists, however, remained invisible; both ghostwriters and secretaries were paid, in the final analysis, not so much for producing a text as for not signing the ones that they *did* produce.[4]

Instruction manuals of the time, such as M. Mostyn Bird's *Woman at Work*, conduct an interesting shell game, depicting women authors who write their own creative work by hand ("scribbling off a short story or article") as overprivileged dilettantes entirely reliant on typewriting agencies to copy and proof their work for them. Until the advent of the typewriter, writing by hand was the pre-eminent sign of authenticity, but here, it stands for the opposite. As Price observes, "The same reversal that reduces the woman who reports her own experience to a fraud transfers authenticity to the woman who copies another's words."[5] In this milieu, Sholes's daughter is every bit as docile as one of Milton's own, an ideal, empty vessel for her father's dictations and her husband's name.

The most famous of Milton's typing great-granddaughters was Theodora Bosanquet. A self-taught typist, Bosanquet was Henry James's secretary from 1907 to 1916, making the same twenty-five shillings a week as any other Type-Writer Girl. Whether James found anything salacious about Bosanquet is a moot point; while he referred to her as "boyish," there is enough speculation about James's sexuality to wonder if that signalled attraction or the lack thereof. James preferred typists that were "without a mind," presumably to maximize the illusion of a transparent dictation. Another of James's typists, Mary Weld, once remarked that "when working I was just part of the machinery."[6] But Bosanquet was far from mindless. An early feminist, she produced the first account of the process of type-writing from the perspective of the female amanuensis, *Henry James at Work*,[7] as well as several other literary biographies, and her diaries are a primary resource for James critics.

Before Bosanquet actually saw James, she encountered his disembodied dictating voice: "My ears were struck by the astonishing sound of passages from *The Ambassadors* being dictated to a young typist . . . I turned round to watch the operator ticking off sentences which seemed to be at least as much of a surprise to her as they were to me."[8] Bosanquet learned that James was about to return from Italy, required an amanuensis, and that "the lady at the typewriter was making acquaintance with his style."[9] Eager to be interviewed, Bosanquet began practising on a Remington typewriter – James was very particular about the brand of his machine. By the time of the interview, Bosanquet wrote, "I could tap out paragraphs from *The Ambassadors* at quite a fair speed." James, however, "asked no questions at the interview about my speed on a typewriter or about anything else." Before long, James requested that Bosanquet find "a small, very cheap *and* very clean *furnished* flat" to serve as "a seat and a temple for the Remington and its priestess."[10]

In *Understanding Media*, Marshall McLuhan uses James and his Remington Priestess as one of his primary examples of the salutary effects of typewriting on literature. According to McLuhan, by the time James hired Bosanquet, typewriting had already become a "confirmed habit" of his, and contributed to the development of a "new style" with "a sort of free, incantatory quality."[11] Moreover, James found typewriting "more inspiring" than writing longhand. James told Bosanquet that "it all seems to be so much more effectively and unceasingly *pulled* out of me in speech than in writing."[12] It's clear from the context it was *speaking to a typist* rather than speech in general that pulled "it" out of James and on to the page. For Bosanquet, the experience of being dictated to was decidedly erotic: "Once I was seated opposite to him, the strong, slow stream of his deliberate speech played over me without ceasing."[13]

As exhilarating as Bosanquet found it, though, the process of typewritten dictation was also difficult:

> The business of acting as a medium between the spoken and the typewritten word was at first as alarming as it was fascinating. The most handsome and expensive typewriters exercise as vicious an influence as any others over the spelling of the operator, and the new pattern of a Remington machine which I found installed offered a few additional problems. But Henry James's patience during my struggles with that baffling mechanism was unfailing – he watched me helplessly, for he was one of the few men without the smallest pretension to the understanding of a machine – and he was as easy to spell from as an open dictionary. The experience of years had evidently taught him that it was not safe to leave any word of more than one syllable to luck. He took

> pains to pronounce every pronounceable letter, he always spelt out words which the ear might confuse with others, and he never left a single punctuation mark unuttered, except sometimes that necessary point, the full stop. Occasionally, in a low "aside" he would interject a few words for the enlightenment of the amanuensis . . .[14]

Both the amanuensis and the master were forced to make accommodations to the "vicious" logic of the machine. Even James's speech patterns adapted themselves to the necessity of typewriting.

Bosanquet confirmed James's suspicions that the use of the typewriter had a definite effect on his style. The only time he resorted to handwriting was when he had to produce something short; typewriting inevitably produced an excess of text. When James allowed himself to dictate short stories from his handwritten drafts, for example, he ended up expanding them "to an extent which inevitably defeated his original purpose."[15] Moreover, when caught up in the throes of dictation, James rather than Bosanquet became the puppet of the machine: "At such times he was beyond the reach of irrelevant sounds or sights. Hosts of cats – a tribe he usually routed with shouts of execration – might wail outside the window, phalanxes of motor-cars bearing dreaded visitors might hoot at the door. He heard nothing of them. The only thing that could arrest his progress was the escape of the word he wanted to use."[16]

McLuhan goes so far as to suggest that toward the end of his life, when James was unable to write without typewriting, *the operating typewriter became the precondition for inspiration*: "He became so attached to the sound of his typewriter operating that on his deathbed, Henry James called for his Remington to be worked near his bedside."[17] By this point, James's ability to write depended not simply on typewriting in general, but on the audio feedback provided by a particular make and model. Bosanquet writes:

> When I began to work for [James], he had reached a stage at which the click of a Remington machine acted as a positive spur. He found it more difficult to compose to the music of any other make. During a fortnight when the Remington was out of order he dictated to an Oliver typewriter with evident discomfort, and he found it almost impossibly disconcerting to speak to something that made no responsive sound at all.[18]

Fleet reporters would later make the same complaint about the Remington Noiseless typewriter.

Critic and biographer Leon Edel confirms McLuhan's contentions in "The Deathbed Notes of Henry James," an article concerning the "Napoleonic fragment" which James dictated during his final days. James suffered a small stroke on December 2, 1915, which confined him to bed, and he contracted pneumonia around December 10. Bosanquet told Edel that "the sound of the familiar machine, and the ability to ease his mind, had helped soothe the novelist in his feverish moments."[19] Six days after his stroke, James called for his typewriter, and dictated a number of fragments over the following days. On December 12, things took a turn for the unusual. James called for Bosanquet and the typewriter before lunch and dictated several short passages, including one whose opening line was "apparently an allusion to his motoring days with Edith Wharton" (significant in this context because Wharton, a female writer, was in the habit of dictating to a *male* secretary using a typewriter[20]). After lunch, James called for Bosanquet and dictated, in a clear and coherent manner, two letters from Napoleon Bonaparte to one of his married sisters.[21] As Kittler observes, "Napoleon's noted ability to dictate seven letters simultaneously produced the modern general staff"[22]; James' ability to channel the ghost of Napoleon during typewriting is one of the symptoms of the emergence of the modern writer's relationship with a haunted technology.

Whether these passages are the result of delirium or James's ability to recall text from volumes of Napoleonic lore in his library is open to conjecture (Edel refrains from labelling James "mad" at this point, but Kittler is less inclined to be charitable[23]).

What interests me is that typewriting creates a situation where pieces of dead and living souls combine and re-combine as the text flows from the typewriter. The dictating voice fluctuates between James and Napoleon; the amanuensis is sometimes Bosanquet, sometimes Peggy James; the addressee sometimes Napoleon's relatives, sometimes James's dead brother William, sometimes Bosanquet, and sometimes posterity. Typewriting not only allows the dead to speak, it allows the living to speak with the dead. The tradeoff, though, is that it becomes impossible to determine the typewritten text's origins.

Bosanquet herself authored several books beyond her memoir of working with James . . . but the dictating voice of James never entirely disappeared from her writing. Henry James died in 1916, but Bosanquet continued to receive dictation from his ghost for decades afterward. She had become interested in the paranormal while James was still alive, discussing the subject at length with her fellow enthusiast, James's brother William, the eminent psychologist and philosopher.

In "Henry James and Theodora Bosanquet: On the Typewriter, in the Cage, at the Ouija Board," Pamela Thurschwell observes that during the 1930s and '40s, Bosanquet attended many seances and practised automatic writing as many as three or four times a day. On February 15, 1933, Bosanquet anonymously attended a seance with a medium named Hester Dowden, who specialized in receiving messages from deceased writers. Via the Ouija board – which, it could be argued, is nothing less than a kind of primitive index-style typewriter designed for receiving dictation from beyond the grave – and a spirit guide named Johannes, Dowden apparently contacted the ghost of James, who was able to answer specific questions about James's private life, including the name of the gardener at his house in Rye.[24]

Moreover, James's ghost, along with the ghosts of George Meredith, Thomas Hardy, and John Galsworthy, was apparently interested in retaining Bosanquet as amanuensis (though Mrs. Dowden, a professional medium, and, after all, knowledgeable in such matters, expressed her scepticism that Galsworthy's ghost would *really* be interested in doing more writing so quickly after his recent death).[25] Several weeks later at another seance, when asked what the plan for proceeding would be, James's ghost responded: "Just to produce an instrument and when that instrument is as efficient as my secretary's typewriter with either a short tale or an essay whichever pleases her best."[26] To the dead as well as the living, then, the Type-Writer (machine) and typewriter (amanuensis) are often indistinguishable.

Bosanquet soon stopped attending Dowden's seances and began transcribing ghostly messages from both Henry and William James. Oddly, she also continued to receive messages from Dowden's spirit guide Johannes, who sometimes acted as an intermediary for the other spirits, and sometimes openly conflicted with them in terms of what he expected. Johannes told Bosanquet that "we want you to make yourself a little more like a nun for the rest of your life on earth because you have to be the instrument of a great work of wonderful and appealing beauty and you must be the first to realise that this work will need all your faculties to make it a little bit of what we want it to be."[27] William James's ghost countered: "We are the last people in the world to make you nunlike or make you take any vow of chastity or of the least."[28] Even for the dead, the body of the Type-Writer Girl remains a combat zone for the forces of business and sexuality.

What did Bosanquet herself hope to gain from all of this spirit-typing? She was after nothing less than recognition from James's ghost of her own status as an author. While James was alive, she repeatedly wrote in her diaries of her desire to live by writing. After James's death, Bosanquet told Johannes that she still wished to be known as the author of "a good book or a good play."[29] Thurschwell argues convincingly that Bosanquet's "fantasy of recognition" culminated in the ghost of James reading her monograph about him, *Henry James at Work*. The spirits tell her that James "thought you were a very uninteresting young woman who had a marvellous gift for transcribing his words correctly, but he . . . finds now that all the time you were observing his style and taking mental notes and that afterwards you wrote a little book about him which he has never had the courage to look at, but he thinks he will have to now if you don't mind."[30] Bosanquet becomes interesting to James's ghost precisely because she has become like him through absorbing his style . . . and because she retains the ability to type for him long after he's turned to ectoplasm. In the case of Henry James and Theodora Bosanquet, the typewriter's triumph is that it asserts the need to be operated even before dictation can begin, and continues to produce writing long after it absorbs dictator and amanuensis alike.

The ghosts of Henry James and Napoleon are far from the strangest entities ever to dictate to a typist. The archive of typewriting is full to bursting with accounts of aliens, ghosts, animals, monsters, insects, and machines expressing strange desires through the machine, and, in many accounts, even becoming the objects of the typist's obscure desires. In the work of William S. Burroughs, typewriting represents both the most repressive and inhuman forms of control imaginable, and the opportunity for escape from that control.

Chapter 13

Burroughs

When he was eighteen years old, William Seward Burroughs (b. 1857), the grandfather of his namesake, William S. Burroughs the writer, became a bank clerk at the Cayuga County National Bank in Auburn, New York. In the late 1870s, the profession of clerking was still entirely manual: rows of clerks at counters and high stools penned in rows and columns of numbers in ledgers in red and black ink. It was monotonous, debilitating work, and Burroughs's contracting tuberculosis didn't help matters.

In 1882 he left the bank and moved to St. Louis, determined to improve the clerk's sorry lot by producing "a machine that could, with one stroke of the handle, add a column of figures and print the operation on paper at the same time."[1] Soon after, he met a machinist and inventor named Joseph Boyer, who rented space and equipment to Burroughs and became an investor in the "Arithmometer" that Burroughs was designing. Four years later, the American Arithmometer Company became a reality, and two years after that, in 1888, Burroughs received the first U.S. patent for a key-operated recording and adding machine.[2]

Burroughs faced considerable scepticism from the financial community, who dubbed him "the mad mechanic." It didn't help that the first run of fifty machines that he produced in 1889 to mollify his investors didn't work; the machines calculated different sums depending on the force with which the handle was pulled. The entire run was recalled, and Burroughs teetered on the edge of failure until he incorporated a hydraulic "dash-pot" to ensure that any amount of force applied to the handle was translated into a uniform amount. By 1891, the machines worked perfectly. Burroughs added a number of other innovations over the years, including duplicate copies on paper rolls, and the reverse mechanism that later found its way into the mechanism of typewriters as well.[3]

The Burroughs company's success also led it, briefly, into the typewriter manufacturing market, thanks to their acquisition of the Moon-Hopkins hybrid adding machine/typewriter. Burroughs produced a manual typewriter in 1931 and an electric model in 1932. Although they sold "a respectable number" of these machines, the company eventually discontinued the production of typewriters and returned to concentrating on the adding machine market.[4]

When William Seward Burroughs was sitting alone in the dark with his fifty recalled adding machine prototypes, he likely would have taken cold comfort in the notion that he had invented the first Futurist writing machine. In the 1914 Italian Futurist manifesto "Weights, Measures and Prices of Artistic Genius," Bruno Corradini and Emilio Settimelli write the following:

> OBSERVATION NO. 2. There is no essential difference between a human brain and a machine. It is mechanically more complicated, that is all. For example, a typewriter is a primitive organism governed by a logic imposed on it by its construction. It reasons thus: if one key is pressed it must write in lower case; if the shift and another key are pressed, it follows that it must write in upper case; when the space-bar is

> pressed, it must go back. For a typewriter to have its E pressed and to write an X would be nonsensical. A broken key is an attack of violent insanity.[5]

Though the Futurists could easily conceive of typewriting as a body-machine assemblage, they would not entertain the notion of a writing assemblage in which they participated but did not entirely control, or what that participation would mean, for better or worse. (They were wrong about the implications of an E key producing X, too . . . as long as the occurrence of this phenomenon is mathematically consistent, it's not insanity but cryptography. I'll return to this point in the last part of this book.)

William Seward Burroughs's grandson – William S. Burroughs the writer – could do all that and much more. And the senior Burroughs would have been entirely surprised at the career that his namesake made of exploring the writer's relationship to the machines that he briefly manufactured. As Burroughs critic Gérard-Georges Lemaire put it, "The writing machine, or typewriter, mythological since Hemingway – who did not realize that the instantaneous nature of the articulated keyboard he used was going to play the same role the camera did in pictorial architectonics . . . was destined to function in an almost autonomous and subversive way."[6]

The Soft Typewriter

With a family tree entwined so explicitly with the history of the technology of typewriting, it's not surprising that William S. Burroughs uses the typewriter as a metaphor for God.

Two of Burroughs's best known novels, *Nova Express* and *The Ticket That Exploded*, refer to an entity Burroughs calls the Soft Typewriter – "the Control Machine"[7] that writes people into being and regulates their behaviour. In *Nova Express*, the phrase occurs twice,[8] but Burroughs explains its operations in detail in *The Ticket That Exploded*. In this book, people are literally typewritten into existence; it describes the human body as "composed of thin transparent sheets on which is written the action from birth to death – Written on 'the soft typewriter' before birth composed of thin transparent sheets on which is written the action from birth to death – Written on 'the soft typewriter' before birth."[9] Burroughs's Soft Typewriter is no less powerful than poet Christopher Smart's typesetter-God. It exercises a ruthless regime of Control (Burroughs capitalizes the word throughout his writing as though referring to an agency or organization) over the bodies that it creates. For Burroughs, Control determines every action throughout a person's life.

It is Burroughs's avowed intention to break the Soft Typewriter's control of humanity by using its own tools – the tools of typewriting – against it. The following passage, from Burroughs's dream journals, depicts Burroughs, in his frequent guise as physician, diagnosing the human condition for a less enlightened soul:

> It's like a typewriter attached to your throat, all the time draining it out of you, you gotta have it, and it is taking more and more . . . *Clinic* . . . *Clinic* . . . *Clinic* . . .
>
> So, Anatole Broyard,[10] you wish to contact me? I think I have half a page for the message.
>
> Please, please, Mr. Burroughs.
>
> Go on, please.
>
> I am told what I will write.
>
> I understand perfectly. Sources?
>
> Don't know. Voices in my head.
>
> Ever try disobeying?
>
> Yes. Results were horrific, headaches and bad luck. Can you help me, Burroughs?
>
> I don't know. I will do what I can.[11]

Describing the amanuensis as the abject slave of a dictating voice is not new; it's explicit in the word's very etymology. In Burroughs's version, the amanuensis, wired directly into the typewriter, has become a junk-sick cyborg powering an assemblage that uses human bodies as batteries as well as components.

As was the case with the assortment of spirit voices that dictate to Theodora Bosanquet and Mina Harker, for Burroughs, the dictator's voice has multiplied into a chorus that comes from an "outside" that may also be inside the writer's head. In Burroughs, typewriting is both alien and haunted, and the characters in his pages are never too surprised when the "Typewriter starts growling like a nasty spirit."[12] The solution that Burroughs proposes is a predictable rhetorical question: "Why not rewrite the message on 'the soft typewriter'?"[13]

Chapter 14

Nothing Is True; Everything Is Permitted

Like typewriting in *Dracula*, the importance of the typewriter to Burroughs's work has been overlooked because it is so utterly central. These kinds of blindnesses are quite common in attempts to describe the systems that we use to inscribe our lives.

Henry Petroski outlines an interesting parallel at the opening of *The Pencil*. In drafting a list of essential supplies for a twelve-day trip into the woods, Henry David Thoreau neglects to include the pencil with which he composed the list, and conducted his extensive journal-keeping . . . even though he had worked with his father at Thoreau & Company, manufacturers of the best lead pencils available in the United States in the 1840s. "Perhaps the very object with which he may have been drafting his list was too close to him, too familiar a part of his own everyday outfit, too integral a part of his livelihood, too common a thing for him to think to mention," writes Petroski.[1] For William S. Burroughs, typing is no longer merely a necessary skill for writers; writing *is* typing: "Sinclair Lewis said: 'If you want to be a writer, learn to type.' This advice is scarcely necessary now. So then sit down at your typewriter and write."[2]

Cut-Ups Self-Explained

Writing is fifty years behind painting. I propose to apply the painters' techniques to writing; things as simple and immediate as collage or montage. Cut right through the pages of any book or newsprint... lengthwise, for example, and shuffle the columns of text. Put them together at hazard and read the newly constituted message. Do it for yourself. Use any system which suggests itself to you. Take your own words or the words said to be "the very own words" of anyone else living or dead. You'll soon see that words don't belong to anyone. Words have a vitality of their own and you or anybody can make them gush into action.

The permutated poems set the words spinning off on their own; echoing out as the words of a potent phrase are permutated into an expanding ripple of meanings which they did not seem to be capable of when they were struck and then stuck into that phrase.

The poets are supposed to liberate the words - not to chain them in phrases. Who told poets they were supposed to think? Poets are meant to sing and to make words sing. Poets have no words "of their very own." Writers don't own their words. Since when do words belong to anybody. "Your very own words," indeed ! And who are you?

CUT THE TEXT INTO THREE COLUMNS:

A	B	C
Writing is fifty y	ears behind painting.	I propose to apply
the painters' techniq	ues to writing; things	as simple and
immediate as collage	or montage. Cut right	through the pages
of any book or newspr	int... lengthwise, for	example, and shuffle
the columns of text.	Put them together at	hazard and read the
newly constituted mes	sage. Do it for yours	elf. Use any system
which suggests itself	to you. Take your ow	n words or the words
said to be "the very	own words" of anyone e	lse living or dead.
You'll soon see that	words don't belong to	anyone. Words have
a vitality of their o	wn and you or anybody	can make them gush
into action.		
The permutated po	ems set the words spin	ning off on their
own; echoing out as t	he words of a potent p	hrase are permutat-
ed into an expanding	ripple of meaning whi	ch they did not seem
to be capable of when	they were struck and	then stuck into that
phrase.		
The poets are sup	posed to liberate the	words - not to chain
them in phrases. Who	told poets they were	supposed to think?
Poets are meant to si	ng and to make words s	ing. Poets have
no words "of their v	ery own." Writers don	't own their words.
Since when do words b	elong to anybody. "Yo	ur very own words,"
indeed ! And who are	you?	

(The letters struck out were those sliced by my scissors. Now, permutate the columns to form the new texts.)

Burroughs's instructions for the cut-up process: propaganda for the war against the Soft Typewriter.

Even though Burroughs sees himself as a product of the Soft Typewriter and very much under its Control, he also recognizes the possibility of rebellion. The principal tool that Burroughs uses to modify his typewriting for insurrection is an invention of the Dadaists called the cut-up.

The cut-up technique first appeared in Dada founder Tristan Tzara's "To Make A Dadaist Poem":

> Take a newspaper.
> Take some scissors.
> Choose from this paper an article of the length you want to make your poem.
> Cut out the article.
> Next carefully cut out each of the words that makes up this article and put them all in a bag.
> Shake gently.
> Next take out each cutting one after the other.
> Copy conscientiously in the order in which they left the bag.
> The poem will resemble you.
> And there you are – an infinitely original author of charming sensibility, even though unappreciated by the vulgar herd.[3]

Like Dada itself, the cut-up creates work that challenges all of the classical assumptions about authorship. The cut-up process came to Burroughs's attention via his friend the artist and writer Brion Gysin. As a precocious teen, Gysin had briefly been part of the Surrealist circle in 1934, until he suffered the same fate as many of the other gay members of that group: excommunication by André Breton, the Pope of Surrealism. But Gysin took the knowledge of Tzara's cut-up technique away with him[4] and, while, in Burroughs's words, "[Breton] grounded the cut-ups on the Freudian couch," Gysin began to conduct his own cut-up experiments and to draw his own conclusions as to their significance.[5]

In Morocco in 1959, Gysin conducted the cut-up experiment that Burroughs cites as ground zero for his own practice, by inadvertently slicing a number of newspaper articles into sections while preparing a mount for a drawing, and then deliberately rearranging them in a random fashion.[6] Before long, Burroughs was using newspaper cut-ups (and eventually, his own typewritten manuscripts) as source material, first typing it out, then cutting it up and retyping it.[7] For Burroughs, the cut-up process became as inseparable from writing as typing: "All writing is in fact cut-ups."[8] Gysin was also experimenting with tape-recorded audio cut-ups, but was particularly impressed with Burroughs's production of "brand new words never said" through typewriter cut-ups.[9] When playing back his early tape cut-ups, Gysin described what he heard as the "squealing of typewriter voices rising out of a Baghdad crowd."[10]

Tristan Tzara wasn't Gysin's only inspiration for this new form of cut-up. While in Morocco, Gysin had become fascinated with the legend of Hassan-I-Sabbah, the founder of the cult of the assassins. The assassin credo "Nothing is true; everything is permitted" echoes like a refrain throughout both Gysin's and Burroughs's work. Gysin relates the following anecdote: early in his career, I-Sabbah received a court appointment to direct the finances of the nation. When the time came for him to present his first report, "his manuscripts had been cut in such a way that he didn't at first realize that they had been sliced right down the middle and repasted . . . All of his material had been cut up by some unknown enemy and his speech . . . was greeted with howls of laughter and utter disgrace and he was thrown out of the administration."[11] From this example, Gysin took the notion of cut-up as a subversion of all forms of bureaucracy and systems of control.

Gysin also took the Arabic word *mektoub* ("It is written") as a literal truth, and reasoned that the cut-up process might allow for a magical manipulation of reality: "If you want to challenge and change fate . . . cut up words. Make them a new world."[12] If the primary purpose of the control that typewriting imposes is to make the most efficient use of office workers' time, it's significant that the disruptions the cut-up creates are also temporal. "Perhaps events are pre-written and pre-recorded and when you cut word lines the future leaks out," muses Burroughs.[13] The metaphor he uses to describe the leakages of time is of typewriting technology gone awry: in *Naked Lunch*, "time jump like a broken typewriter"[14] and, many years later, in *The Place of Dead Roads*, "time jumps like a broken typewriter."[15]

Burroughs came to think about his typewriter cut-ups as a collaboration "on an unprecedented scale" with an indefinite number of writers, both living and dead.[16] Like Theodora Bosanquet, Burroughs describes his own work as a collaboration with dead writers: "While I was writing *The Place of Dead Roads*, I felt in spiritual contact with the late English writer Denton Welch, and modeled the novel's hero, Kim Carson, directly on him. Whole sections came to me as if dictated, like table-tapping."[17] Using the cut-up method puts Burroughs in touch with even more dictating voices. In "It Belongs to the Cucumbers," Burroughs's most detailed statement on his cut-up process, he notes that experimenting with tapes invariably produces voices not present on the original recordings. Burroughs suggests several "outside" sources for these voices, the voices of the dead, including the voice of Goethe, among them (though, with characteristic sardonic humour, he notes that "many of them have undergone a marked deterioration of their mental and artistic faculties"[18]).

Burroughs is also willing to entertain other possibilities, including extraterrestrial sources ("no reason to think we have a monopoly on banality") and the "backplay of recordings stored in the memory banks of the experimenters" – that is, a kind of leakage of the writer's subconscious onto the writing material.[19] Ever the pragmatist, Burroughs simply accepts the presence of these voices as given and considers the matter of their origin "an academic question as long as there is no way to prove or disprove it."[20]

So what, exactly, did Gysin and Burroughs use the cut-up to change?

Queer Typewriting

After his 1959 cut-up experiments, Gysin "realized right away that cut-ups would never serve or suit anyone quite like they fitted William and served him."[21] First, Burroughs had a massive amount of typewritten material on hand to slice and recombine – the legendary "word hoard," a trunk-filling megamanuscript that was the source for *Naked Lunch*, *Nova Express*, *The Soft Machine*, and *The Ticket That Exploded*. But Gysin had a sense that the cut-up was a way to insert theretofore-excluded subjects into the discursive register of the day. "[I]t was all red hot stuff about firsthand reports on areas not even spoken about let alone written about and printed, except in the closet. Burroughs arrived fresh on the scene with heroin and homosexuality . . ."[22] Typewriting's willingness to connect with anyone and anything was a major factor in the entry of a multiplicity of possible sexualities into mass culture.

Something else that allowed Burroughs to type the atypical was his publisher. Maurice Girodias, the publisher of Olympia Press, created an environment at his publishing house where "for the very first time a modern writer didn't have to sit down in front of his typewriter and say 'Oh gee, I guess I can't write this, 'cause I'll never get it published.' "[23] Whenever Gysin writes of Girodias, he describes him as yet another addition to the technologies surrounding the typewriter: "[Girodias] opened the *language box*, man, he opened the *typewriter*, he opened the *minds* of all the American writers who came to Paris and found it agreeable to stay here, and realized that for the first time they didn't have to put on their own *self-censor* when they sat down in front of a piece of paper."[24]

Despite the charges of obscenity that have been levelled at his work over the years, the sexuality that Burroughs delineates through typewriting is not intentionally pornographic in the conventional

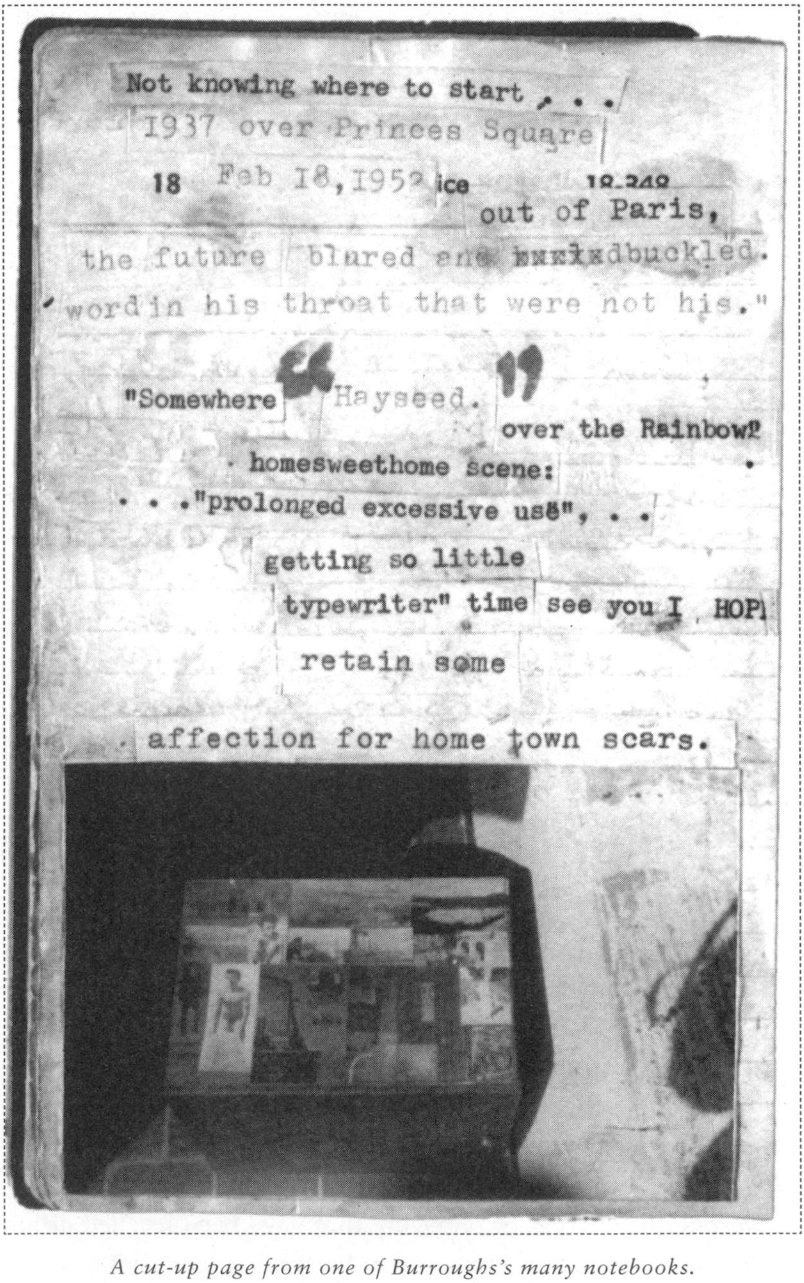

Not knowing where to start . . .

1937 over Princes Square

18 Feb 18, 1952 ice 18,348

out of Paris,

the future blured and buckled.

word in his throat that were not his."

"Somewhere "Hayseed." over the Rainbow?

homesweethome scene:

. . . "prolonged excessive use", . .

getting so little

typewriter" time see you I HOP

retain some

affection for home town scars.

A cut-up page from one of Burroughs's many notebooks.

sense. What Burroughs is after is a careful description of a sexuality that, until he sat down to write it himself, he could not find in other written accounts. Burroughs's typewriting was capable of articulating queer desires, in a style and language that left no doubt that what was being articulated was different from what had gone before it.

For Burroughs, cutting up texts is itself a sexual act.[25] Burroughs's typewriting describes a liquid world where words and solid objects – including typewriters, of course – merge in unorthodox couplings, beyond all attempts to control them, as in this description of cruising gay bars: "Here where flesh circulates in a neon haze and identity tags are guarded by electric dogs sniffing quivering excuse for being – The assassins wait broken into scanning pattern of legs smile and drink – Unaware of The Vagrant Ball Player pant smell running in liquid typewriter."[26]

The only thing that sex in Burroughs's writing is not is *romantic*. Emotional, definitely, but for Burroughs, "love is a fraud perpetrated by the female sex" and "the point of sexual relations between men is nothing that we could call love."[27] His admitted misogyny – Burroughs describes women as a "basic mistake" and part of the order that he wishes to cut up[28] – means that as revolutionary as it was in terms of articulating queer desire for a mass audience, his typewriting also threatened to erase many of the gains made by Type-Writer Girls. This troublesome facet of Burroughs's work is inextricable from the progressive part, and it persists, virally, into the work of another artist who has documented some of the more sticky and disturbing aspects of typewriting in the contemporary imagination: David Cronenberg.

Chapter 15

Channelling Burroughs

David Cronenberg's *Naked Lunch*[1]: the title of the 1991 feature-length film adaptation of Burroughs's most famous book reads like a cut-up text itself. When describing the relationship that his film bears to Burroughs's writing, Cronenberg echoed the sentiments of the other writers and artists I have described who felt they were receiving dictation strong enough that it could continue beyond the grave: "I started to write Burroughsian stuff, and almost felt for a moment, 'Well, if Burroughs dies, I'll write his next book.' Really not possible or true. But for that heady moment, when I transcribed word for word a sentence of description of the giant centipede, and then continued on with the next sentence to describe the scene in what I felt was a sentence Burroughs himself could have written, that was a fusion."[2] It should come as no shock, then, that the symbol Cronenberg chose to depict that relationship is the typewriter.

One of the richest and most interesting threads running through Cronenberg's *Naked Lunch* is the one dealing with the complexities of typewriting. In the film, a bewildering variety of part-mechanical, part-biological typewriters become metaphors for the varying and often contradictory mental states of the film's characters, almost all of them writers. Cronenberg is quite clear that the film's focus on typewriters is not from Burroughs's writing, but from his own obsession with artists' tools in general and writing machines in particular.[3] When Cronenberg began writing the script for his *Naked Lunch*, while living in England in 1989, his inspiration for the insect typewriters came from observing the insects attracted to the glowing screen of his Toshiba laptop at night. Cronenberg notes that once he began the treatment, it was almost as if his laptop was writing the script itself.[4]

What eventually emerged from Cronenberg's laptop was an imaginative and controversial depiction of the relationship between typewriting and a much older technology: drugs.

Poison and Cure

Drugs, writes Avital Ronell, are "technology's intimate other."[5] Intimately linked with dictation, drugs create "a secret communications network" with the writer's internal/external dictating voices; "something is beaming out signals."[6]

As Marcus Boon has detailed in *The Road of Excess: A History of Writers on Drugs*,[7] drugs play a major role in Burroughs's writing. In the first paragraph of his Introduction to *Naked Lunch*, Burroughs writes that he has "no precise memory of writing the notes which have now been published under the title *Naked Lunch*."[8] For the author, the entire book is a dictation received from elsewhere.

Burroughs's description of the relationship between drugs and writing suggests that narcotics prepared him to act as an amanuensis: "Under morphine one can edit, type, and organize material effectively but since the drug acts to decrease awareness the creative factor is dimmed."[9] Brion Gysin describes Burroughs's writing process during this period as a kind of receptive passivity: "When he found himself in front of the wrecked typewriter, he hammered out new stuff. There were already dozens of variants, and, if something seemed missing, slices of earlier writing slid silently into place alongside later routines because none of the pages was numbered."[10]

This is not to say that Burroughs advocated narcotics as a tool for writers. Cannabinoids are the only class of drug that Burroughs credited as being of any active use for his typewriting process; "passages written with cannabis have stood the test of critical after-inspection" ("amphetamines and cocaine are quite worthless for writing and nothing of value remains" and "I have never been able to write a line under the influence of alcohol").[11] Large sections of *Naked Lunch* were written in Tangier, when Burroughs had kicked his junk habit but was using several forms of cannabinoids; "much of the atmosphere of *Naked Lunch*, the turbulent blocks of association and intense paranoia, along with the constant shifting between hard-boiled pulp realism and experimental dream writing, have more in common with cannabis literature than with anything written about narcotics."[12] And, to be entirely fair, Burroughs himself noted that his writing also depended on being *off* drugs: "*Naked Lunch* would never have been written without Doctor Dent's [apomorphine] treatment."[13]

The drugs in Cronenberg's *Naked Lunch* are of an entirely different order: "bug powder," "the black meat of the aquatic centipede," and "mugwump jism" are all non-existent substances that represent the ways that society's institutions coerce people into living according to their regulations. The physical "agents" of Control in the film are all giant insects – an irrational representation of the alleged forces of rationality. Ground and processed, their bodies provide the drugs that the writers in the film constantly ingest. Left intact, they serve as the typewriters on which the writers in the film compose. William Lee, the film's protagonist and Burroughs stand-in, has a highly ambivalent relationship with these insects. As a part-time exterminator, he is opposed to them from the outset; as a writer and an eventual addict, he needs them as much as he loathes them.

It is not until after Lee shoots his wife, Joan, that the film actually depicts him typing. Until this point, it has not been either necessary or possible because both Cronenberg's film and Burroughs's own writing position a spousal murder as the moment where authorship begins.[14] Lee pawns the gun with which he shoots Joan for the "Clark Nova" typewriter that allows him to write.

When Lee injects "the black meat of the aquatic centipede" in front of this typewriter,[15] two things occur. The first is that the typewriter begins to operate as if manipulated by some unseen force. The second is that the typewriter transforms into a giant insect, with a talking sphincter in the middle of its back (an allusion to Burroughs's infamous "talking asshole" routine from *Naked Lunch*) . . . and a typewriter keyboard on the portion of its head between its mandibles.

The bug-typewriter is Cronenberg's solution to an otherwise unfilmable moment. He observes that "the act of writing is not very interesting cinematically. It's a guy, sitting . . . basically he sits and types. It's an interior act. In order to really convey the experience of writing to someone who hasn't written, you have to be outrageous. You have to turn it inside out and make it physical and exterior."[16] On the level of plot and motivation, the asshole typewriter becomes the vehicle that allows Lee to write unspeakable truths, even ones he doesn't want to write.[17] Its anal mouth simultaneously equates truth with shit (as the first half of the Hassan-I-Sabbah mantra claims, "Nothing is true") and speaks the queer desires that Lee himself is unwilling to articulate from his own mouth (as the second half of Hassan-I-Sabbah's mantra claims, "Everything is permitted").

When Clark Nova speaks, the dictatorial logic of typewriting asserts itself with a vengeance. One of the first things it says to Lee is, "I want you to type a few words into me. Words that I'll dictate to you." Lee's laconic but cautious response is, "Sure, what the hell."[18] Clark Nova goes on to inform Lee that the words it is dictating come from a nebulous organization known only as Control;

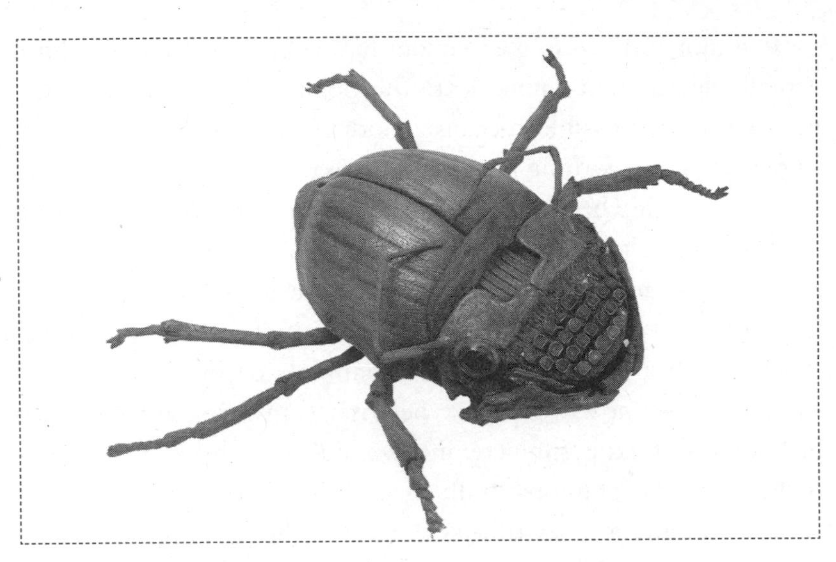

The Clark Nova, William Lee's "all American" bugwriter.

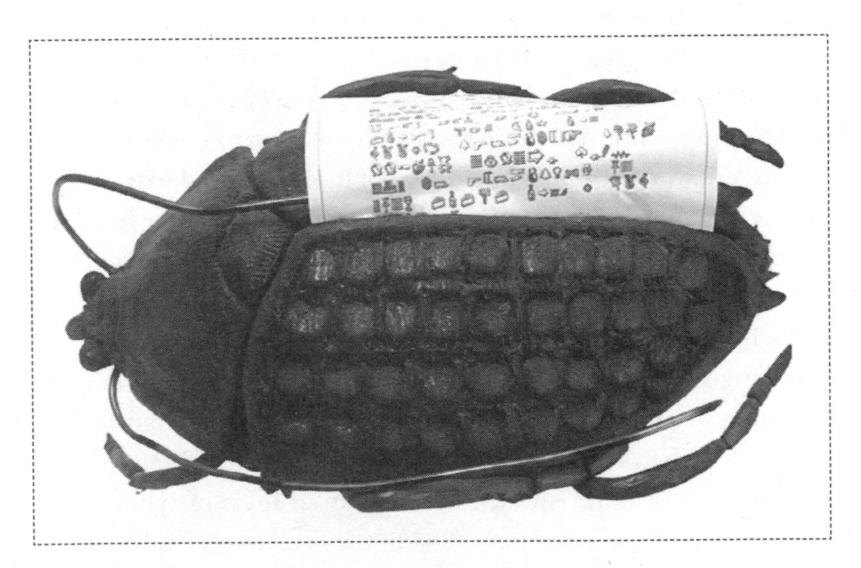

An early prototype for the Clark Nova, not used in the film.

whether Control has Lee's best interests at heart is a moot point, even if it is a symbol of Lee's own internal muses.

Notably, the insect typewriter does not permit or articulate *all* desires. Lee can *speak* the phrase "I was a homosexual" while on a date with another man, but as he nods out over his "report," there is a blank space in his typewritten page where the phrase "I was a homosexual" should appear.[19] The line "shall never forget the unspeakable horror that froze the lymph in my glands when the baneful word seared my reeling brain:" ends with a colon and a space, a typographic reminder of both the truth that Lee cannot bring himself to write and the asshole (colon) in the back of the typewriter that will not write it for him. This refusal of articulation will drive Lee further into paranoia and cause much more damage before the film is over.

The writing machines in the film function increasingly as manifestations of the mindsets of their owners and operators. As Lee's world becomes more fragmented, he appears with multiple typewriters on the table in front of him[20] (this scene also alludes to Burroughs's many descriptions of his multiple tape-recorder setups for subverting systems of control through the cut-up process[21]). As he ventures out, Lee encounters cafés full of people typing,[22] and typewriters begin to act as links between two writers, transforming the ambivalence Lee feels toward his own writing machine into genuine antagonism.

Shortly after discovering the typing café, Lee experiences a very real "return of the repressed" when he meets Joan Frost, an American expat writer whose husband, Tom Frost, is also a writer. To risk a particularly Cronenbergian pun, Joan Frost is a dead ringer for Joan Lee, Bill's murdered wife; Judy Davis plays both characters.[23] When Lee encounters the Frosts, they are both writing, but with a difference: Tom writes with a typewriter, and Joan writes longhand. When Lee remarks on this in a later scene, she says, "I'm not good with machines – they intimidate me."[24]

In the terms of the film, writing longhand is hardly writing at all; real writers use typewriters. In his director's notes, Cronenberg explicitly identifies the crippling of a writing machine as a blow to the potency of the writer[25]; Tom Frost remarks at another point that "I feel desperately insecure without a typewriter in the house."[26] When Tom Frost later kidnaps Lee's Clark Nova in retaliation for the destruction of his Martinelli, and the insect-machine demands "a full report" as retaliation, Lee sneers, "How am I going to write it? Longhand?"[27] When Tom offers to loan Lee his Martinelli typewriter, he says "you could borrow her,"[28] simultaneously attributing a female identity to his writing machine and creating an ambiguity that echoes the ambiguity that has always surrounded the word "typewriter": is he offering the loan of his writing machine, or of Joan? Lee eventually "borrows" and destroys both.

Friedrich Kittler argues at length that it was the typewriter that freed women from passively acting as either the object of writing or the tool used for that process of investigation. Cronenberg's *Naked Lunch*, on the other hand, suggests that typewriting can also reverse women's emancipation. Clark Nova informs Lee that not only was he "programmed" to shoot his wife Joan Lee, but that he was justified in the act because she wasn't human, "or, more precisely . . . [was] a different species from men." According to Lee's typewriter, Joan is an enemy agent, an inhuman "elite corps centipede" who married him as her cover, unaware that he was the counter-insurgent agent assigned to her case.[29] This becomes Lee's alibi not only for writing the unthinkable, but for murder itself.

"It's Arabic"

On that note, the third typewriter in Cronenberg's *Naked Lunch* – the Mujahideen – enters the film during Lee's seduction of Joan Frost. When Lee asks about its origins, Joan responds vaguely that "It's Arabic."[30] The Mujahideen is the epitome of what Edward Said describes as "Orientalism" in his classic book of the same name: a colonial will to understand, control, and manipulate elements of non-Western cultures as "the Orient." Orientalism has much less to do with a well-defined geographical area or specific culture than it has to do with the ways that Western culture constructs images of the East as an inversion of itself.[31] In effect, "the Orient" is a mirage: what the West *wants* to see rather than what is.

Said's particular definition of the term is "the Anglo-French-American experience of the Arabs and Islam, which for almost a thousand years stood for the Orient."[32] This is precisely the backdrop for Cronenberg's *Naked Lunch*: an imaginary city called Interzone – Burroughs's term for Tangier, when it was still a free city and an "International Zone" – that is a kind of superposition of Morocco onto Manhattan (at various moments in the film, one can view minarets out the window from scenes set in New York, or Central Park from scenes set in Tangier). The experience of William Lee and the Frosts, or of Burroughs, Paul and Jane Bowles, and Brion Gysin in Morocco, is the experience that an expatriate has of a colonized country, that is, an experience that sees only what it is convenient to see. Heightening the effect is the fact that the entire Interzone set was not filmed on location, but in a warehouse in Toronto.

The Mujahideen typewriter itself is interesting for a variety of reasons. Like any mirror image, it is reversed on the horizontal axis, and indeed the typewriter produces Arabic script, moving from right to left across the platen as it is operated. The following passage from *The Story of the Typewriter* indicates that the mechanics of Arabic script present convincing reasons for the choice of an Arabic typewriter to represent an "Oriental" logic contrary to typewriters that produce monospaced Roman script:

> Of all the languages now written on the typewriter, the Arabic group presented the gravest mechanical difficulties. The Arabic character, as written, is not subject to any of the usual rules. It has in its complete alphabet over one hundred

> individual characters; it writes backwards, i.e. from right to left; the characters are written on the line, above the line and below the line, and they are of various widths, requiring full spacing, half spacing and no spacing at all.[33]

Like all narratives, *The Story of the Typewriter* has an agenda. If the notion of constructing an Arabic typewriter presents a "medley of problems well calculated to tax ingenuity to the limit," the text raises these difficulties in order to present another moment of victorious re-territorialization for the logic of typewriting: "the Arabic typewriter is a crowning triumph of mechanical skill."[34] The existence of an Arabic typewriter, in these terms, represents the triumph of Western reason and technology over a set of rules so different that they are classified as "alien" in the strongest sense of the term.

The design of the Mujahideen typewriter is not even based on any typewriter actually produced in an Arab country. In his director's commentary, Cronenberg misstates that the Mujahideen's design is based on an English typewriter from the World War Two era[35] when

The Mujahadeen typewriter, in mid-transformation from machine to Sex Blob.

it is clearly based on the Oliver typewriter, invented by an Iowa minister in 1888. Oliver typewriters are immediately identifiable because they feature concave type bars divided into two even groups, on either side of the machine, that resemble bellows; the type bars strike obliquely down toward the machine's centre at the platen. The Oliver company ceased production in 1928, but was extremely successful, and produced over a million machines during the thirty-four years of its existence.[36]

The machine takes its name from the Arabic word for "strugglers" (*Mujahid*: someone who exerts effort or struggles). It refers generally to Islamic guerrilla warriors, specifically to those armed and trained by the United States, Pakistan, and Saudi Arabia to fight the Soviet invasion of Afghanistan between 1979 and 1989. Since the Soviet withdrawal from Afghanistan, the term has also come to refer to those resistance fighters that have opposed American troops, first in Kuwait during the Gulf War of 1990–91, and, at the time of this writing, in Iraq circa 2004. (Many Muslims did in fact volunteer to fight in Afghanistan during the Soviet era in order to gain experience as guerrilla warriors, including fighters associated with Osama Bin Laden.[37])

The choice of "Mujahideen" resonates in several important ways as a name for this machine. First, as insurgents trained and armed by an imperial power that they later turned upon, the Mujahideen are roughly analogous to the precarious "double agent" status that Lee himself possesses in relation to the insect typewriters and their superiors.

Second, Cronenberg's original plan to film *Naked Lunch* in Tangier was scuttled by the first Gulf War, resulting in the literal creation of an Orientalist Interzone in a Toronto warehouse. As Soviet imperial adventures in Afghanistan created the first Mujahideen, U.S. imperial adventures in Kuwait, Iraq, and Afghanistan created their Orientalist double, not merely the typewriter, but the entire physical milieu in which it appears.

Third, there are at least two levels of misrecognition in Cronenberg's statements about the origin of the Mujahideen typewriter. Presumably, he selected it as an emblem of one of the imperial powers with a vested interest in Tangier. However, the specific materiality of the machine was not particularly important to Cronenberg, who selected it for its "provocative" design in order to serve as a "strange sexual vehicle" for the ménage of Joan, Lee, and the typewriter. In this respect the typewriter functions in a very similar manner to the young male hustlers of the Interzone medina, who serve as intermediaries for and triangulate the desires of various expatriate writer couples in the film.

Lee's seduction of Joan is the only typewriting scene in the film in which dictating voice, writing machine, and amanuensis are clearly separable entities (and even then, the number of dictating voices is multiple). When asked by Joan if he intends to kill Tom's Mujahideen as well as the Martinelli, Lee replies, "Only in self-defense. I understood writing to be dangerous. I didn't realize the danger came from the machinery." Lee convinces Joan, against her better instincts, to use the black meat,[38] then the typewriter, while he crouches behind her and whispers in her ear. Though he can't read what she is typing, as it is in Arabic script, he goads her to write "more erotic," then "filthier," then "unthinkable." He begins, quietly, to whisper in her ear, passages from Burroughs's *Naked Lunch*.

The mechanics of dictation in this scene are quite complex. Lee is receiving orders from Control. Joan, who is, according to Clark Nova, an agent of Interzone, Inc., is creating an erotic text, but we have no idea who or what is dictating it to her. Lee then begins to dictate from the very novel that inspired the film in which he is appearing, a film that the director claims he has, in essence, channelled from his inspiration. The symbolic manifestation of all of these conflicting desires and voices is the transformation of the Mujahideen into a "Sex Blob": the keyboard goes soft, then folds into a large, mucosal organ; the keys turn into bellows, which begin to sigh; a phallus emerges from the rear of the machine, which then transforms completely into a headless, tentacled torso that wriggles and flops onto the two writers, who have collapsed in ecstasy on the floor.

What ensues is a combination of the realization of a Tijuana Bible Type-Writer Girl fantasy and a soft-porn version of Burroughs's own writing, where there is a direct connection between sexuality and machines: the machines add an element of "impersonal cruelty."[39] There's a lot of gasping and thrashing around, but no climax, as the woman Lee later resentfully identifies in a typed report as "the bitch queen Fadela," a "lesbian agent of Interzone, Inc.," interrupts and chases the Sex Blob over the balcony with her riding crop; it crashes to the pavement as nothing more than a broken typewriter. If the scene had reached its logical conclusion, there would be some basis for arguing that typewriting in the film created or authorized desires outside of heterosexuality or a misogynistic version of homosexuality, but the Sex Blob only turns out to be an elaborate form of tacky sex toy. So much for the revolutionary power of the Mujahideen typewriter; once Western writing is finished with it, it becomes no more than another object to be salvaged by some nostalgic journalist on his way home.

The Last Writing Machine

In retaliation for the destruction of his Martinelli, Tom Frost kidnaps Lee's Clark Nova, and the intoxicated Lee takes to wandering through the Interzone medina with a pillowcase full of the "the remains of my last writing machine." When the audience sees the contents of the sack, it contains not the broken insect parts of the Martinelli, which Lee scooped out of his bathtub, but a collection of pills, pipes, needles, and other drug paraphernalia. (In his director's commentary, Cronenberg draws a parallel between drug paraphernalia and typewriters[40] as the tools of particular kinds of obsessive subcultures.) At his most abject point, Lee is rescued a second time, by Kiki, a street hustler. He takes Lee to a machine shop where the fragments of the Martinelli are recast into the film's final biological typewriter – the Mugwriter. The aged Burroughs himself was the physical model for the Mugwumps in the film[41]; here, finally, is a symbol of the process of inspiration and dictation that Cronenberg describes as his impetus for writing the film in the first place: channelling Burroughs.

After the artisan pulls the white-hot Mugwriter out of the forge, the film cuts to Lee typing on the keys in the head's open mouth, drinking "Mugwump jism" from a mug that collects the oozings of the phallic growths on the Mugwriter's head. The link between writing, drugs, and queer sexuality becomes ever stronger, and the typing assemblage ever more compact. But the paranoia continues; after the Mugwriter tries to dupe Lee into a potentially fatal situation, Lee takes it to Tom Frost for "an exchange of hostages" in the hopes of recovering Clark Nova (even within its case, the Mugwriter continues to spout muffled but clearly elaborate conspiracy theories). Frost returns the battered remains of Clark Nova to Lee on the grounds that it's "too damn All-American" for his use; Lee resounds that "the Mugwriter is so foreign that it's almost alien," pointing once again to a kind of Orientalism at work. What constitutes "foreign" and "All-American" is not only in flux, but has little to do with any genuine alterity that the citizens of Interzone might possess. What matters is what the expatriate writers choose to see as same or other at any given instant.

In this final encounter with Frost, Lee's first transaction is reversed: along with the soon-to-expire Clark Nova, he receives a gun in exchange for the Mugwriter. The circle of the film is drawing to a close; as he did at the opening, Lee will once again shoot Joan, but willfully this time, as a demonstration of his ability to write. "I can't write without her," he says to Benway, as he rescues Joan from the Interzone, Inc. drug factory. A more unflinching response, the kind Lee alleges that he makes in his reports, would have been

"I can't write without her dead." In the discursive universe of Cronenberg's *Naked Lunch*, operating a typewriter and operating a pistol are equivalent acts ("We could use a man of your caliber – .32, isn't it?" mocks Benway),[42] and otherness, especially the otherness of women and foreign boys, is the target. "All agents defect and all resistors sell out," Clark Nova tells Lee with its last breath[43]; political concerns beyond the raw exigencies of survival are, to paraphrase Dr. Benway, "unfortunate side-effects."[44] Type-Writer Girls and Boys looking for salvation through writing would do well to look elsewhere, because the period when they were the privileged operators of writing machines may have already drawn to a close.

Part

Authority:
Typewriting as Discipline

"The words that you have so painfully organized are swiftly and precisely organized for you by this machine."

– August Dvorak, *Typewriting Behavior*

Chapter 16

Pen Slavery

An odd metaphor recurs in many of the documents from the early era of typewriting. *The Story of the Typewriter* assures its readers repeatedly that the typewriter "freed the world from pen slavery"[1] or "the bondage of the pen."[2] Before typewriting, the pen symbolizes all of the phallic power that is associated with authorship. It is the sign of order and the production of law, and of the powerful and wise individual who wields it.

Even under the regime of typewriting, and, later, networked word processing, the handwritten signature still retains its position as the sign of power, authority, and personal assent. The first document to actually use the word "typewriting," the *Scientific American* editorial on Pratt's Pterotype, assumes that even the promise of the typewriter will have to bend to this necessity: "The weary process of learning penmanship in schools will be reduced to the acquirement of the writing of one's own signature."[3] Christopher Latham Sholes allegedly abandoned the pen completely after he built his first working typewriter, typing even his signature. However, even the usually worshipful author of *The Story of the Typewriter* considers this act "extreme, even by [the standards of] the present-day business man."[4] The answer to the problem of who is being freed from what, exactly, must exist elsewhere.

"Pen slavery" actually expresses the frustration of a bureaucracy, in both the public and private sectors, crying out for a more efficient tool to manage the mountains of paper it generates. The nineteenth-century wave of globalization that was British colonialism created, among other unforeseen difficulties, unprecedented logistical challenges. George Nathaniel Curzon, during his tenure as Viceroy of India, epitomizes the sentiment with his weary exclamation that "the tyranny of the executive pales into insignificance when compared to the tyranny of the pen."[5] "Freeing the world from pen slavery" turns out to involve finding some other sort of more efficient writing mechanism with which to enslave the world in the name of capital.

In a world where bureaucracy functions according to a pen-based regime, the height of efficiency is a well-managed room full of copyists, equipped with an array of pens, coloured inks, blotters, carbon paper, and double-entry ledgers . . . which is the same as saying that it's not very efficient at all. The tedium of this type of work was significant enough that the figure of the Dickensian clerk, with his fingerless gloves, wire-rimmed spectacles, and eyeshade, perched on a tall stool and hunched over his copy desk, remains a cultural cliché to this day.

From the popular fiction of the time, it's also evident that the distribution of power in the pen-based clerical system had its limits. As if to usher in the period of the greatest number of attempts to invent a writing machine, Herman Melville published the short story "Bartleby the Scrivener" in 1853. Bartleby, a copy clerk whose quiet but insistent demurral "I would prefer not to" first exempts him from proofing his own copy work, then from labour altogether, nearly becomes the ruin of his employer, who actually has to move his place of business to get away from his passive-resisting erstwhile employee. Though a general level of cultural anxiety about Bartleby-like figures may have helped to fuel the invention of the modern typewriter, typewriting did little to address the problem. Bartleby, living, eating, and sleeping behind his screen in the corner of his office, is the archetype of the cubicle-dwelling dot-com hacker/slacker employee circa 1997.

What was required was not only a writing machine that enabled quick, accurate reproduction of multiple copies of documents, but a kind of cultural logic that would provide instructions about how to train bodies to use the new writing machine effectively. It would have to provide some clues as to how to reorganize those bodies in multiple sets, according to its own logic, to scale pen-based offices up to the level where they could accommodate the kinds of organizing that business on a global scale demanded. And it would have to produce writing that was clear, verifiable, and replicable, writing that had the status of truth.

Martin Heidegger wrote (with a pen, one assumes) that "the typewriter tears writing from the essential realm of the hand . . . The word itself turns into something 'typed.'"[6] Typewriting, in other words, rewrites first language itself, then the body of the typist, then of the world around the typist, in its own image.

Amaranath sasesusos Oronoco initiation

Typewriting places an invisible grid onto the blank page: one character or space per cell, no more, no less (at least, until IBM announced the invention of proportional spacing in 1941[7]; even after this point, many typewriters in production were single-spaced).

Single-spacing is in fact a step backward from the sophistication of typesetting technology, but for many years it was necessary for the typewriter to function. While marshalling printed text into the cells of its disciplinary grid, the typewriter exerts its power on more than the letters, but also on the way that words convey meaning.

Consider the sentence "Amaranath sasesusos Oronoco initiation secedes Uruguay Philadelphia." The meaning of this sentence has nothing to do with the normal logic of syntax and everything to do with the logic of how the letters appear on a typewritten page. It was usually the first thing ever typed on each new typewriter, and its sole function was to check the alignment of a typewriter that had just rolled off the production line before it was shipped. Unlike most sentences, it was rarely spoken, and no one particularly cared what it might mean in the conventional sense.[8]

How does it work? "Amaranath," the misspelled name of an imaginary flower, checks the alignment of the vowel "a" between a number of common consonants. "Oronoco" checks the "o" key, while "secedes," "initiation" and "Uruguay" check three vowels that are among the most commonly used of all letters, "e," "i," and "u." "Sasesusos" not only compares four of the five vowels in the same word against the baseline of the letter "s," but also "includes several of the most common letter combinations in twentieth-century business English."[9] "Philadelphia" checks the horizontal alignment of "i" and "l," the narrowest letters on the keyboard.

Instead of conveying meaning, typewriting reconfigures words and sentences as tools that serve as an affirmation of the machine's own precision.

Rewriting Control as Writing

Between the time that Christopher Latham Sholes received his first patent for a typewriter (ca. 1870) and the years between the two World Wars (ca. 1930) the process of representing the world through art, especially literature, became industrialized along with everything else. Creating images and texts was no longer the sole privilege of artists and writers of refined sensibility. It became a process with definite steps, a process that could be reproduced on the assembly line. And, for the first time, the representation of work in writing, and all of the attendant filing, sorting, processing, and publishing, became an equally important part of the work itself.[10]

We ourselves live so thoroughly inside of this mentality that it's difficult to conceive of any work in terms other than programming, sorting, reproducing, and circulating documents and images of someone doing something else.[11] In the most famous soliloquy from Cameron Crowe's film *Say Anything*, everydude Lloyd Dobler summarizes this difficulty in a manner reminiscent of Bartleby: "I don't want to sell anything, buy anything or process anything as a career. I don't want to sell anything bought or processed or buy anything sold or processed or process anything sold, bought or processed, or repair anything sold, bought or processed. You know, as a career, I don't want to do that."[12] Lloyd would never have had to worry about such things if it were not for typewriting, which played an important role in turning writing into work and work into writing. What typewriting contributed to this process was a way of reorganizing not just the letters on the page, but also the bodies and minds of typists.

Bodies in Space

When we think of power in the abstract, we tend to think of it as something monolithic, oppressive, and negative that is wielded over us by Those In Charge. But there are other kinds of power that sometimes escape our notice. They emerged during the seventeenth and eighteenth centuries, and have persisted to the present day. These exercise themselves through schools, hospitals, prisons, and other social institutions. We expose ourselves to them voluntarily, and they modify our attitudes and everyday behaviour in ways that make us more productive citizens[13]: they make us literate, they teach us to drive, they inoculate us to prevent outbreaks of illness, and they teach us to type, among other things. From a cynical perspective, the attractiveness of these forms of power for heads of state, as opposed to techniques of domination and slavery, is that they are *cost-efficient*: producing a useful population is at least as successful a form of control as trying to violently force a population to behave in a certain way (if not more so), and does not require all of the investments that appropriating the bodies of others by force entails.[14]

By the seventeenth and eighteenth centuries, armies, schools, and hospitals already had developed techniques like drilling and timetabling that produced bodies that were useful to them. Advances occurring at the same time in anatomical studies and philosophy were producing techniques that made it easier to understand the operation of body and mind. But there was also a crossover point, epitomized in Julien Offray de La Mettrie's essay *Man a Machine*,[15] describing a body that was both manipulable and intelligible, a body that power could transform, improve, and use for its own ends. The writing automata that preceded the typewriter represented the kinds of bodies these technologies of power wished to produce: clockwork people who would flawlessly and agreeably complete any assigned task.[16] (This dream has persisted, along with the forms of power that make it possible, though as movies like *I, Robot* demonstrate, we're now much more worried about the potential rebellion of our automata than we used to be.)

The techniques used by these new technologies of power differed in important ways from the oppressive and violent forms of power that preceded them. First, the scale of control that they exercised was completely different. It focused on micromovements, gestures, and the relative speeds with which these were accomplished. Second, these new forms of power were not as interested in how bodies looked from the outside as in such matters as how their component parts were organized, or how to improve the efficiency of their movement. Third, the new power was interested in process rather than results, and thus exercised itself continually and constantly, for example, choosing to incarcerate and rehabilitate a criminal rather than occasionally and violently publicly torturing and executing a criminal. The name for this new form of power is "discipline."[17]

"Discipline," Michel Foucault writes, "proceeds from the distribution of individuals in space."[18] It operates by dividing the world up into grids: spaces designed to contain and manage different segments of a population. The obvious examples are monasteries, barracks, and prisons, but in the case of typewriting, we could also include factories and trade schools. Discipline partitions the world into ever-smaller spaces in order to wring maximum productivity out of each individual, according to their various skill sets[19]: welcome to the cubicle. This kind of organization ensures that all bodies are effectively interchangeable, because relative positions and abstract relationships are what's important in the system, rather than specific individuals.[20] In other words, it doesn't really matter who the new VP of Fundraising is, as long as the person in that job understands their relationship to the other VPs, the President, and their various underlings.

Using models devised long ago in monasteries to divide the day for cycles of prayer, discipline partitions time as well as space. From these religious systems come timetables and schedules, which establish the daily, weekly, monthly, and annual rhythms and cycles of modern life. These systems create ever-finer divisions of time that permit the breaking down of any action into sequential steps.[21] The finer the subdivisions of time, the more possibilities emerge for uses of that time that might make work even more efficient.[22]

As a result of discipline's division and mapping of bodies in time and space, any action can be described in terms of the time that it takes to perform (see illustration on page 148).[23] Correlating body parts and timed gestures leads to the notion that there is such a thing as the "correct use" of the body in the performance of any action, and that no movement must be wasted, no fraction of time left unused.[24] This was the birth of the sciences of time and motion studies and ergonomics.

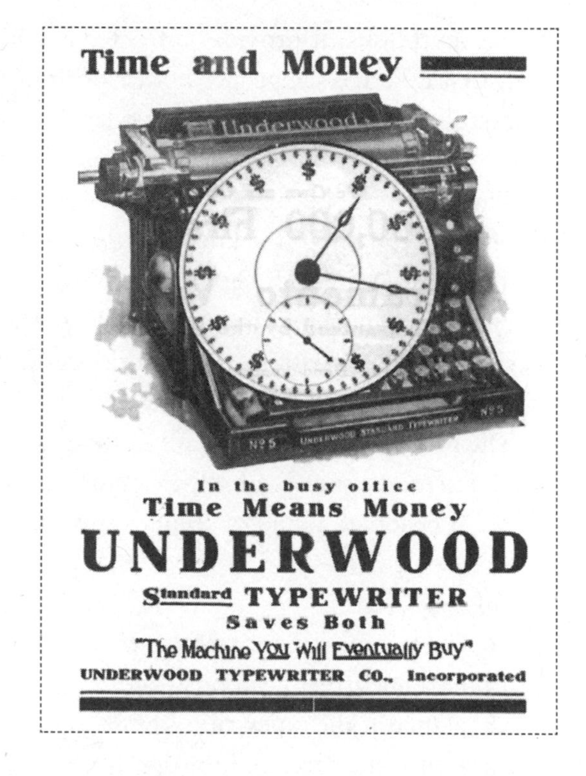

Time and Money
Underwood
In the busy office
Time Means Money
UNDERWOOD
Standard TYPEWRITER
Saves Both
"The Machine You Will Eventually Buy"
UNDERWOOD TYPEWRITER CO., Incorporated

The dollar signs on the clock face in this turn-of-the-century ad make it abundantly clear that the partitioning of workers' time into smaller units via typewriting means more efficency . . . and therefore more profit.

As discipline reclassifies bodies into their component parts, it begins to treat those parts as objects, which means that they can be considered in combination with each other, or with inanimate objects,[25] and makes it possible to conceive of such things as the optimum number of hands (not people) required to operate an assembly line, or a typing pool. A mixture of organic and mechanical components is optimal because meat, muscles, bones, and sinews can interact repetitively with tools and weapons according to tolerances that are not only measurable, but also susceptible to gradual improvement, thanks to the application of exercise.[26]

Nowhere does this system of discipline become more clear than in the standard classroom typing manual. Helen Mogyorody's *Typing 100*,[27] a standard choice for high school, business school, and college courses for most of the 1970s, is exemplary. Like most typing textbooks, it provides an extensive list of precise instructions that describe the proper bodily positions during typing, such as the following, the third item on a list of ten:

> There should be approximately 6 to 8 inches of space between the top of the knee and the frame of the typewriter. The front of the body should be from 8 to 10 inches from the base of the machine. The front of the frame of the typewriter should be even with the edge of the desk.[28]

The monastic roots of discipline are nowhere more evident than in the relentless process of self-examination that the text instils. In places, it becomes a kind of secular catechism:

> Do you
> 16. insert your paper carelessly?
> 17. have incorrect head position?
> 18. glance at the typewriter keys?
> 19. return the carriage improperly?
> 20. hesitate after carriage returns?
> 21. jam your keys?
> 22. forget about the right margin or bell?
> 23. have faulty space bar stroking?
> 24. operate the shift key poorly?
> 25. pound the keys?
> 26. have jerky typing?
> 27. hesitate when typing the "number row"?
> 28. neglect to use the machine conveniences?
> 29. worry about making errors?
> 30. freeze to the keys?
> 31. fail to follow directions?
> 32. overestimate your speed skill?
> 33. proofread negligently?

34. have faulty tabulation?
35. lack interest?
For every "YES" answer, promise to improve that fault. If you have NO faults, CONGRATULATIONS![29]

Bless me, Father, for I have neglected to use the machine conveniences.

It is this form of power that infamously enabled Henry Ford to write that of the 7,882 actions involved in creating a Model T, only 949 of them required whole bodies; "670 [of them] could be filled by legless men, 2,637 by one-legged men, two by armless men, 715 by one-armed men and ten by blind men."[30] This is the form of power that lies at the heart of the American Dream, and it is the form of power that laid the ground rules for typewriting.

Chapter 17

Therbligs

Frank Gilbreth was born in 1868 – the same year as Sholes's typewriter – at the dawn of the control revolution, and would live to become one of its chief architects. Fittingly enough, he began his professional life as a bricklayer, then moved up through the ranks to building constructor, before embarking in 1912 on the activities that he helped to transform into professions: motion studies, management consulting, and industrial engineering. What Frank Gilbreth and others like him developed was a way of writing about work that became a form of work itself.

His partner in that career was Lillian Moller, whom he married in 1904. Gilbreth Inc.'s core revelations were a result of Frank's tenure as a bricklayer. While observing the techniques of the other bricklayers at work, Gilbreth realized first that every one of them performed the same task in a slightly different manner, and that on the average it required about eighteen motions to lay a brick. He decided that adjusting the spatial relationships between the bricklayer's body and tools would make the task easier to perform and more efficient overall. Gilbreth designed an adjustable scaffolding with a shelf for bricks and mortar that allowed the bricklayer to pick up a brick with one hand and scoop mortar with the other, without having to stoop for each brick. He hired cheap unskilled labourers to pre-sort the bricks, arranging and packing them with their best faces pointing in the same direction, saving time for the bricklayer, who could confidently position each brick in the emerging wall. By the time Gilbreth was finished with his micromanagement of the bricklayer's physical environment, he concluded that laying in a single brick should only require four and a half motions.[1]

That "half motion" should give us pause. What, exactly, constitutes half a motion, and where, for that matter, does one motion end and another begin? Though Gilbreth and his peers strove to give motion studies the legitimacy of a science, there were definitely some dodgy aspects to it.

From his bricklaying studies, Gilbreth hypothesized the existence of "one best way" to perform any given job, and began to design a set of concepts that would help him discover that one best way for any given task, not just bricklaying. In 1911, Frank and Lillian published their theories in *Motion Study*, a book that attempted to deduce general patterns of movement from observing factory workers in action. With the aid of stopwatches and cinematic and long-exposure photography, the Gilbreths settled on a system of seventeen basic arm and hand activities, broken down and classified in terms of a precisely described and timed (in milliseconds, and sometimes to the microsecond) unit of movement they dubbed a "therblig" (spell it backwards). The theory of motion studies described in their book soon became a bona fide discipline, and the information that it garnered permeated all aspects of work, from the perspective of both management (eager to optimize work areas for greater efficiency) and labour (which utilized motion studies for workplace safety audits and for establishing standards of work rates during contract negotiations).[2]

One of the Gilbreths' first major clients was E. Remington & Sons, and they did much to cement the typewriter into the daily

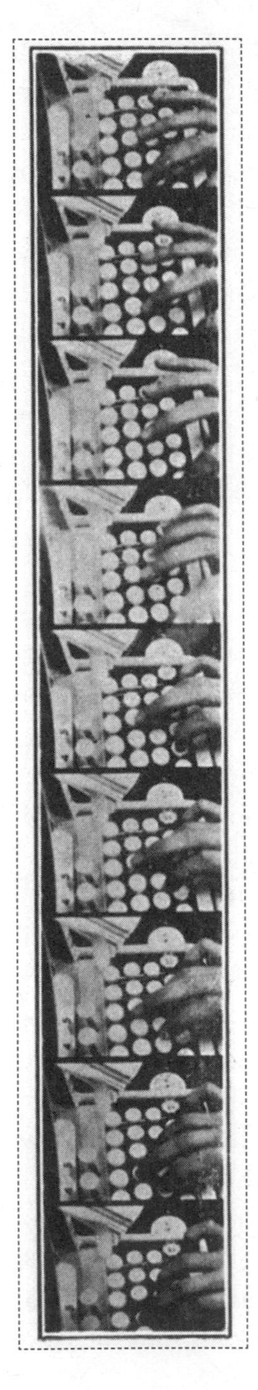

LEFT:
Frank Gilbreth's micromotion study of the fingers of Miss Hortense Stollnitz, International Amateur Typing Champion, performing at her fastest speed.

RIGHT:
Another micromotion study of Miss Stollnitz, changing the paper in her machine.

operations of modern business. But the Gilbreths felt that their theories had applications in more aspects of life than the workplace. They were one of the major forces that began to move the discourse of micromanagement and motion studies out of the office and into the home.

Moby Dick Comes Home

Among their other impressive accomplishments, Frank and Lillian produced twelve children in seventeen years. For the Gilbreths, this meant, among other things, that they had plenty of test subjects for their theories. One of their sons, also named Frank, and his sister Ernestine wrote the bestselling *Cheaper by the Dozen*[3] (which has been made into two films), which relates the origin of domestic engineering in the Gilbreth household: "It was just about impossible to tell where [Frank Gilbreth's] scientific management company ended and his family life began," relates the book on its opening page.[4] Given that its subject matter is a childhood with eleven siblings and parents all too willing to advance "the elimination of wasted motions"[5] – a phrase that begs the question of what, exactly, constitutes a wasted motion in childhood, or, conversely, which excessive motions should be corrected – *Cheaper by the Dozen* is a somewhat rosy account, coloured in Norman Rockwell tones. (Those of a more cynical bent will want to temper their reading of this book with a viewing of *Peeping Tom*, a 1960 *film noir* about a child raised by an efficiency expert; his father's motion study experiments on him drive him to transform the camera his father used to document his life into a murder weapon.[6])

Frank Gilbreth's program for increasing the efficiency of his family depended heavily on the very technologies he himself was helping to refine. Writing was central to this process, as was the use of charts and graphs. "Every child old enough to write – and Dad expected his offspring to start writing at a tender age – was required to initial the charts in the morning after he had brushed his teeth,

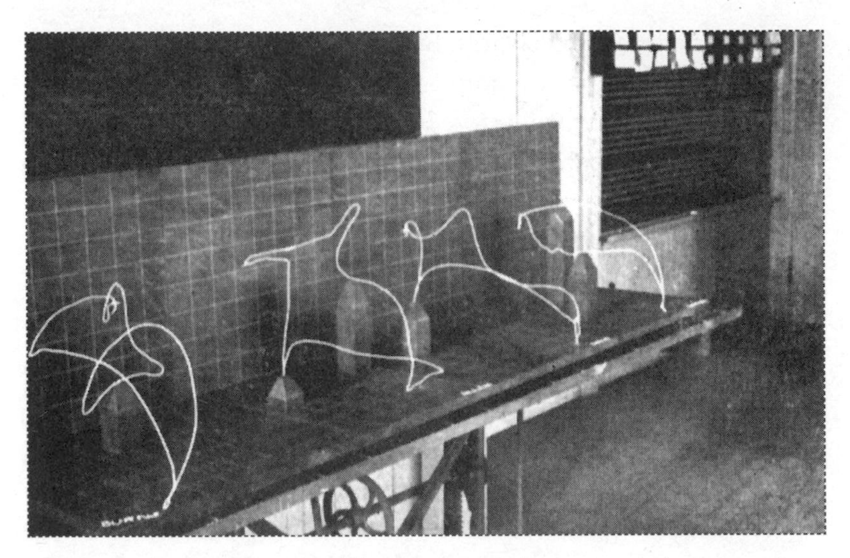

Gilbreth's first photograph of the wire models he constructed to chart, from left to right, the progress of a man learning "paths of least waste" for moving his left hand while operating a drill press. The subject was a manager who hadn't operated the machine in 25 years.

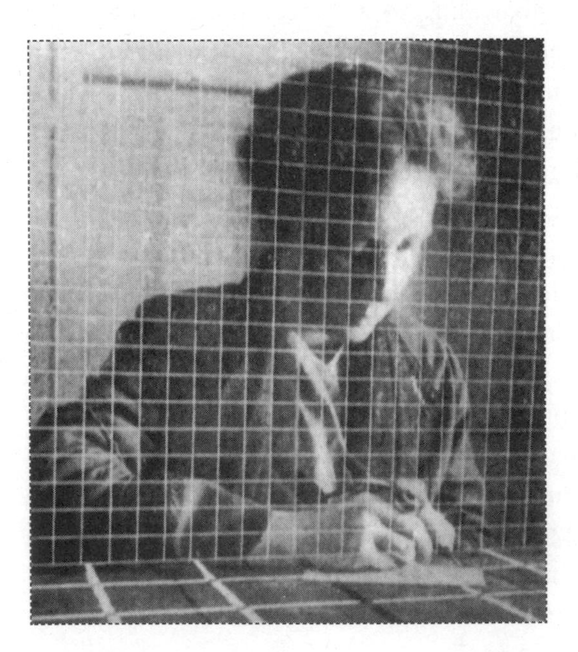

This is the first photograph of the "penetrating screen" Gilbreth used to isolate and analyze the motions of a worker completing a task. The invisible grid of the typewritten page imposes itself on the world.

taken a bath, combed his hair, and made his bed," write Frank Jr. and Ernestine. "At night, each child had to weigh himself, plot his figure on a graph, and initial the process charts again after he had done his homework, washed his hands and face, and brushed his teeth."[7] The terms in this ordering system are microprocesses, disembodied parts, and statistical averages rather than complete acts performed by sovereign individuals; there is no practical distinction between actions performed on inanimate objects and actions performed on body parts. What matters is that all sets of teeth are brushed, all hair is combed, all beds are made. Such a system does more than create order; it also instils the processes of order into those that use these systems as a kind of self-justification. Teaching the Gilbreth children to plot all of this data on themselves was as much or more about instilling the value of the process of plotting data as an ordering device as it was about gathering information, and created, in effect, a dozen tiny middle managers.

When Gilbreth began to consult for Remington, the arsenal of prostheses he employed to create order out of the unruly tangle of his children's bodies expanded to include the typewriter. A scene early in *Cheaper by the Dozen* involves Gilbreth recounting to his family one night over dinner how he had "helped Remington develop the world's fastest typist" by "put[ting] little flashing lights on the fingers of the typist and tak[ing] moving pictures and time exposures to see just what motions she employed and how those motions could be reduced."[8] The choice of the verb "develop" to describe the process of training a touch typist could just as easily describe the development of yet another model of the Remington typewriter itself.

After claiming he has developed a system that can teach anyone to touch type in two weeks, even though he himself is a hunt-and-peck typist (a fact his children find hilarious) the "Great Experiment" of Gilbreth teaching his children to type begins.[9] He brings home an all-white typewriter, which he nicknames "Moby Dick,"[10] explaining, "It's white so that it will photograph better" – an important

The Gilbreth family: cheaper by the dozen.

consideration in a profession where his movement analysis studies are based largely on photographs – and "Also, for some reason, anyone who sees a white typewriter wants to type on it. Don't ask me why. It's psychology."[11] The glib phrase "It's psychology" turns an entire field of discursive practice into a conceptual black box. Want someone to type? Use psychology. Gilbreth might as well have said, "It's magic," because there is no causality here, an odd attitude for a man whose entire professional reputation rides on bloodless rationality – unless his statement is itself part of his overall attempt to make his children desire to use the machine, which he initially refuses to let anyone else touch. In this "optional" two-week experiment, the grand prize is the typewriter itself, with a watch and pen as consolation prizes "awarded on a handicap basis, taking age into consideration."[12] It is a beautifully closed circuit of self-justification: the prize for learning how to typewrite is . . . the typewriter.

Before any of the children are allowed to touch the typewriter, Gilbreth first requires them to memorize the positions of the letters on the keyboard, with the help of a diagram. "QWERTYUIOP . . . Get to know them in your sleep. That's the first step."[13] Next, he colours the fingers of the children with chalk, to match coloured zones on their typewriter keyboard diagram: "For instance, the Q, A, and Z, all of which are hit with the little finger of the left hand, were colored blue to match the blue little finger."[14] The children practised for two days. Ernestine, the fastest typist, won the privilege of sitting down at the white typewriter first, only to discover that her father had placed blank caps he custom-ordered from Remington over all the keys so she can't see what she is typing. Gilbreth instructs her to imagine that the keys were coloured and that she was typing on her diagram. "Ern started slowly, and then picked up speed, as her fingers jumped instinctively from key to key."[15] To this system, Gilbreth adds a straight behaviourist trick, standing over the typing child and hitting her on the head with a pencil every time she hits an incorrect key: "It's meant to hurt. Your head has to teach your fingers not to make mistakes." Within two weeks, all of the children are typists.

Gilbreth photodocuments the whole process, claiming the photos are for his files only, but a month later, they become the substance of a newsreel "which showed everything except the pencil descending on our heads. And some of us today recoil every time we touch the backspace key."[16] By way of justification, the younger Gilbreths write, "Dad had a knack for setting up publicity pictures that tied in with his motion study projects,"[17] but the overall picture, the project of Gilbreth teaching his children to type, is actually about producing documents about teaching people how to type, promoting not simply touch typing, but the discourse of motion studies itself as a form of labour. The convenient omission of the blows with the pencil whitewashes the process to match the typewriter, itself a blank page waiting for whatever impressions the discourse of efficiency might place upon it.

Chapter 18

QWERTY

By imposing an invisible grid, the typewriter remakes in its own image first the page, then the body of the typist, then the world around the typist. The principle begins with the page. In monospaced type, which was the only kind of type for much of the history of typewriting, each letter occupies the same amount of space: one cell in the grid.

The grid moves off the page to affect the body of the typist via the keyboard. Most people already understand this instinctively, because we can recite its visible aspects like a kind of technological catechism: QWERTYUIOP, the order of the top row of keys on most typewriter keyboards. (Let's call it QWERTY for short; everyone does.) Frank Gilbreth exhorted his children to run through the letters like so many prayer beads (the typing exercise may have been voluntary in the Gilbreth household, but for that matter, so was prayer[1]). The logic behind QWERTY is so ingrained, even today, that it is almost invisible to us. When referring to it, many sources even deign to call it "universal," despite plenty of evidence to the contrary. And, as *The Story of the Typewriter* smugly declares, "The universal keyboard has a hold similar to that of language itself."[2]

There are a wide variety of competing explanations for how the configuration of the QWERTY keyboard came about. Barring any one convincing explanation, most have to be relegated to the status of "factoids," a word coined by Norman Mailer in *Marilyn*, his biography of Marilyn Monroe, to designate "an assumption or speculation that is reported and repeated so often that it becomes accepted as fact."[3] Not that their status as factoids makes them any less interesting; factoids can be deployed in various contexts – knowingly or unknowingly, maliciously or innocently – but there is always something at stake (usually the claim of the author to some sort of authority).

The lesson learned from examining these conflicting accounts is that the typewriter, like most technologies, is more of a highly stratified conglomerate of various half-assed solutions and dead ends than the shining culmination of a logical mechanical evolution, where rationality and common sense dominate at every turn.[4] With uncharacteristic frankness, Beeching's *Century of the Typewriter* baldly states that the idea that the arrangement of the keys was in any way "scientific" was "probably one of the biggest confidence tricks of all time."[5]

Today's QWERTY keyboard is fairly close to Sholes's original design. Photographs of the patent diagram for the QWERTY keyboard (U.S. patent no. 207,559) and of the Remington Type-Writer No. 1 show some variations: beginning on the right of the top row are the numerals 2 through 9, hyphen, comma, and em dash; the colon key follows the P on the second row; on the third row, a broken vertical bar precedes the A, and the M follows the L; the fourth row begins with the ampersand; the C and X are transposed; and the N is followed by the question mark, semicolon, period, and apostrophe.[6] But how did it get that way?

Sholes's friend and associate, Dr. Roby, hypothesized that Sholes originally set the keys in alphabetical order, and pointed to the FGHJKL string on the middle row as evidence.[7] Beeching makes a

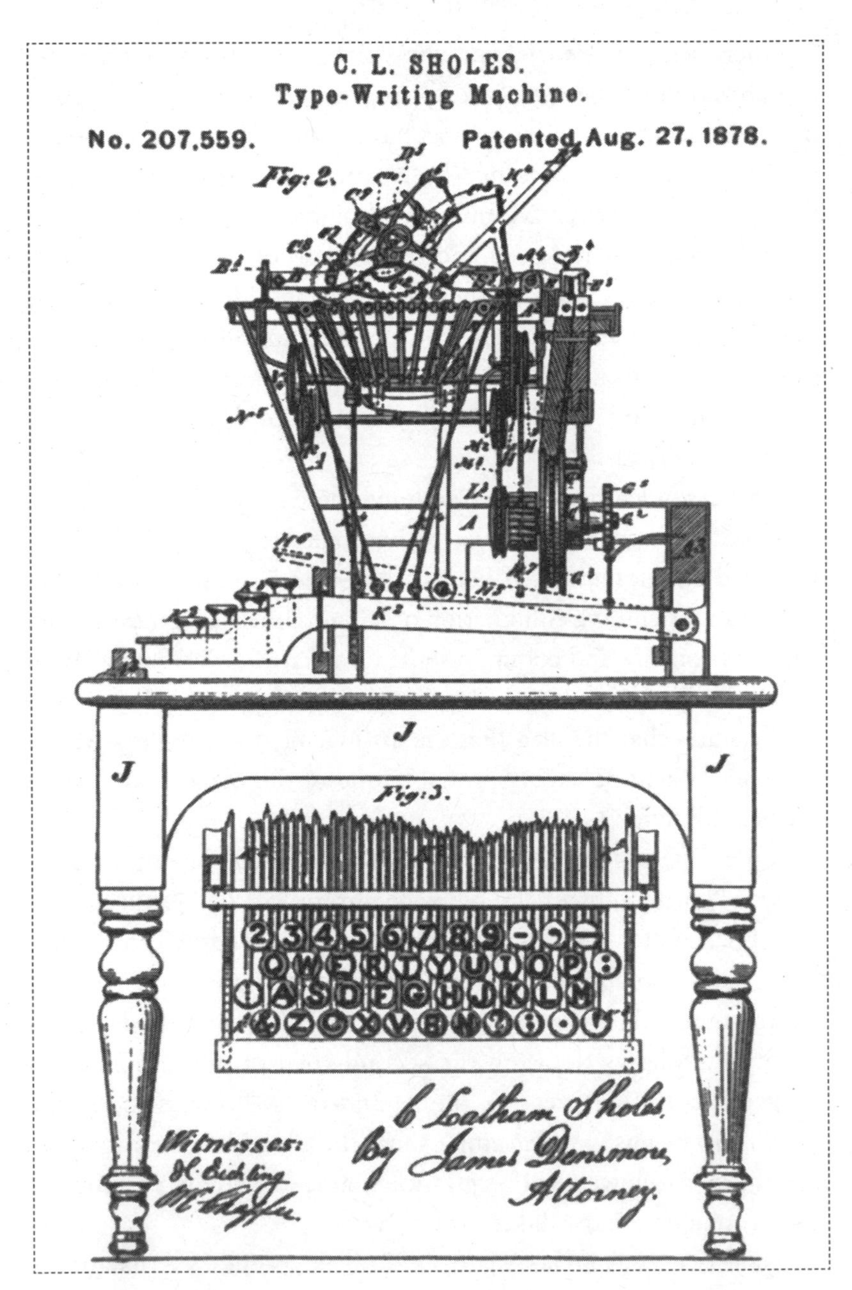

Christopher Latham Sholes's 1878 patent, featuring a QWERTY keyboard nearly identical to the ones still in use today.

similar claim: "Originally the tendency was to arrange all letters . . . in alphabetical order for easy reference. It was assumed that if people knew their alphabet, and most people did, it would be easy to locate the letter required." Perhaps the alphabetical arrangement of keys was actually *more* efficient than the typewriter could handle, because, the argument continues, "the 'ABC' arrangement caused his up-strike machine to jam when any speed was reached."[8] Whether or not this was empirically true (and recent experiments may not support it), it's one of the many competing claims for how the QWERTY keyboard developed.

Torbjörn Lundmark repeats another common factoid about the QWERTY keyboard in *Quirky QWERTY*: even those typewriter salespeople without any actual skill or speed in typing "could easily demonstrate what it did, how it worked – even what it was called – by using their index finger to pick out in letters only from the top row" the word typewriter. Lundmark, however, cannot say whether this happened by accident or by design[9]; so this factoid perpetuates the circulation of a particular statement, and provides Lundmark with a snappy ending to his chapter on "How the QWERTY Keyboard Was Invented" without actually providing any real information.

If Sholes's experience as a printer had any bearing on his keyboard design, it would have been to convince him that an alphabetic arrangement was of little use. Remember that Sholes's original typewriter had only an upper case; the standard pattern (or "lay") for upper case characters in a printer's type case was usually alphabetical order, proceeding from upper left to lower right, in rows of seven cells. That relatively awkward arrangement exists largely by virtue of neglect, because upper case characters are used much less frequently than lower case. The QWERTY layout does not follow the layout of any of the bewildering variety of U.S. printers' lower case type cases of the era.[10] Lower case lays were highly complex affairs that had to take characters such as ligatures (ff, fi, fl, ffi, ffl, and so on) and spaces into account; the only passing resemblance they bear to keyboards is that the most common lower-case configurations have four rows, with the numerals running in a narrow row across the top.

The matter on which all sources seem to agree is that whatever configuration Sholes started with, the type bars began to collide with each other and stick when a typist of even moderate speed began to type.[11] In *Cognitive Aspects of Skilled Typewriting*, William Cooper suggests that one of Sholes's major questions was the problem of how to minimize the jamming of the keys slowly returning to their home position with those just being struck. The solution was to move the keys that commonly stuck together to opposite sides of the keyboard. To this end, Sholes's partner James Densmore asked his son-in-law, the school superintendent for Western Pennsylvania, to prepare a list of the most common two-letter sequences in the English language.[12] Speculation is that Sholes and Densmore then used this list to split up as many of these pairs as they could. Beyond the fact that they requested a table of letter frequencies, this is once again venturing into factoid territory, because no one seems to have a copy of the list or any evidence of how Sholes and Densmore implemented it.

According to Bliven, the net result of Sholes's arrangements is a keyboard "considerably less efficient than if the arrangement had been left to simple chance."[13] Although most typists have more dexterity in their right hands than their left, the QWERTY keyboard has a decidedly sinister bias. The statistics Bliven provides state that the left hand makes 56 per cent of all keystrokes when typing in English, but the little finger of that hand is overworked, being responsible for the two most difficult keys to work on a manual typewriter, the shift-lock and the back-space keys. Forty-eight per cent of all finger motions on the QWERTY keyboard are one-handed, in order to evenly distribute the workload; an optimal number would be no more than 33 per cent.[14]

More than anything else, the QWERTY keyboard is an example of how the arbitrary can become normalized and even lionized. In an age of type balls, daisywheels, and digital character buffers on even the most basic of office machines, to say nothing of word processors, the QWERTY configuration is completely unnecessary, but it still persists.

But what are the alternatives?

Chapter 19

Dvorak (or, In the Navy)

There have been numerous proposals for alternate keyboard configurations over the years, but the most popular is the one August Dvorak and his colleagues developed at the University of Washington in the early 1930s. Dvorak patented it in 1936 as the Dvorak keyboard, or the Dvorak Simplified Keyboard. Cooper's *Cognitive Aspects of Skilled Typewriting* claims that its improvements include "a larger home-[middle] row vocabulary (3000 vs. 100 common words), greater utilization of right-hand keying, more balanced utilization of all fingers of each hand . . . and minimization of awkward fingering sequences." In addition, Cooper writes, a typist can learn the Dvorak keyboard in a third of the time that it takes to learn the QWERTY, and achieve speeds from 15 to 20 per cent higher with 50 per cent fewer errors and less operator fatigue.[1]

So why doesn't everyone use a Dvorak keyboard? Beeching presents a slightly conspiratorial thesis: "Nobody could agree on what a new keyboard should be, but the biggest opposition came from *teachers of typing* as it still does today. They wanted things to remain as they were, and they are still the most reluctant to

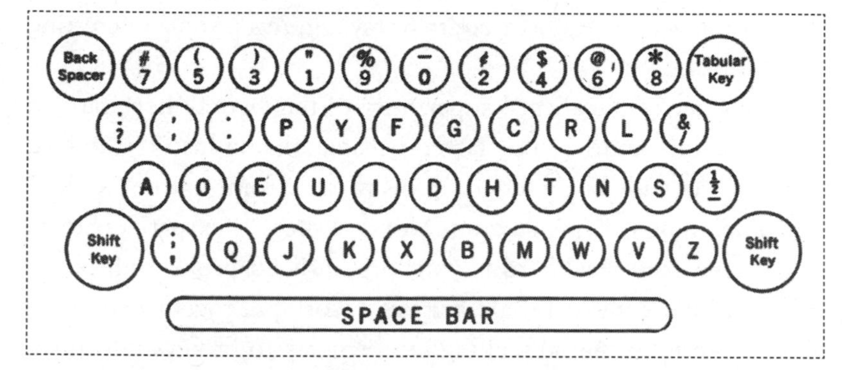

Dvorak's diagram of his "Simplified" typewriter keyboard.

change their methods and learn all over again."[2] Damn those typing teachers and their evil schemes. One can almost see Beeching's lip curl with contempt as he hammers out the phrase "teachers of typing."

Conventional wisdom says that when consumers are presented with a choice between two products, one of which is technologically superior, not always will the superior product become the standard. Even though competition encourages commodification, which drives costs down and spurs innovation, consumers are generally unable to co-ordinate their purchases to the extent that they can act *en masse* in their own interest. Sometimes, they cannot even ensure the continued manufacture of the superior product; sometimes, it is relegated to the status of a boutique item serving a niche market. In either case, the technologically inferior product becomes by default the standard. The popular examples of this phenomenon are Beta videocassette recorders versus VHS recorders, Apple computers versus IBM clones, and, in both the economics literature on the evolution of standards and the popular press on the subject, Dvorak keyboards versus QWERTY keyboards.

Economists S. J. Liebowitz and Stephen E. Margolis are not the kind of people who are satisfied with conventional wisdom. In a very persuasive article titled "The Fable of the Keys," they re-examine the Dvorak versus QWERTY controversy and reach some interesting conclusions.

The story of the alleged superiority of the Dvorak keyboard to QWERTY implies that the latter became the industry standard without ever truly being tested. In his massive tome on the psychology of learning to type, August Dvorak writes, "the spatial pattern of the 'universal' keyboard is your first difficult problem. It is too difficult. This keyboard is a crazy patchwork put together too long ago in a series of heartbreaking experiments to fit keys into positions without their colliding or sticking, and so to invent a usable machine. It was put together for a few fingers at a time when no one dreamed of fast, all finger, touch typing."[3] One could only reach this conclusion by bracketing the hundreds of other typewriter configurations invented over the years and classifying them as irrelevant. At the turn of the nineteenth century and well into the 1920s, typewriter manufacturers often supplied typists along with their machines. Liebowitz and Margolis argue that "since almost every sale required the training of a typist, a typewriter manufacturer that offered a different keyboard was not particularly disadvantaged," and as offering a different configuration was a matter of changing the order of the keys and soldering different slugs onto type bars, the cost of offering an alternate keyboard would not have been prohibitive. Therefore, when typewriter manufacturers began to move toward QWERTY configurations, it could not have been simply because it was already a dominant standard. Likewise, many of the companies that did originally manufacture alternate keyboards – for example, Hammond, Blickensderfer, Yost, and Caligraph – all made reputable machines that contributed important technical refinements, so nonstandard keyboards were not doomed to failure by being inadvertently associated with faulty technology.[4] The reason for QWERTY's dominance must lie elsewhere.

According to Liebowitz and Margolis, the claims for the superiority of the Dvorak keyboard are suspect because of a variety of flaws in both the circumstances and the methodology of the research. The writing of economist Paul David[5] is the point where the narrative of the Dvorak keyboard as an example of the market failing to select a superior standard enters the literature of economics. Like many writers, David refers to "experiments done by the U.S. Navy that had shown that the increased efficiency obtained with the DSK [Dvorak Simplified Keyboard] would amortize the cost of retraining a group of typists within ten days of their subsequent full-time employment" without citing the exact documents. His assertions are repeated in turn in other economics and popular nonfiction accounts (including virtually all of the popular typewriter histories – Bliven's, Beeching's, and even Lundmark's recent addition to the genre) and gain authority in the process.

Arthur Foulke's biography of Christopher Latham Sholes, *Mr. Typewriter*, mentions an Associated Press report dated October 7, 1943, which breezily claimed that the Dvorak keyboard enabled typists to "zip along at 180 words per minute" but also suggests that there has been at least a typographical error, if not embellishment, when he notes that *Business Week* reported the test speed as 108 words per minute on October 16, 1943. Foulke did some fact-checking of his own and was assured by the Navy that it had not conducted such a speed test and had not made an official announcement to that effect.[6]

What the Navy *did* do was conduct four experiments, two that "were not truly fair tests" and have no apparent record of their objectives, methods, or results, and another two conducted in July and October 1944. The July study involved retraining fourteen Navy typists for two hours a day on "overhauled" Dvorak keyboards. The subjects had an average IQ of 98 (the societal average is supposedly 100) and dexterity skills that averaged at 65 per cent (on which scale I have no idea). A dozen of them were already QWERTY typists, but substandard ones with an average speed of 32 words per minute. After spending an average of 83 hours on the

Dvorak keyboard, their average typing speed increased 75 per cent to 56 words a minute.

In the second experiment, eighteen typists whose IQs and dexterity skills are not recorded were retrained on QWERTY keyboards (unlike the first experiment, there's no mention of the condition that these typewriters were in). The average initial speed of these individuals was 29 words a minute, but because three of them started at zero words a minute, their beginning and ending speeds were calculated as averages of their first four and last four tests. As Liebowitz and Margolis point out, this averaging considerably reduced the margin of increase reported for the QWERTY typists, which was a 28 per cent increase to 37 words a minute.

There may also have been other variations from the methodology of the first test, as the report is sketchy, but it's already clear that comparing the two sets of results would be difficult to do fairly. For one thing, there's no way to tell if these results would hold true for average or skilled typists as well as substandard ones. Yet the Navy study concludes that "indisputably, it is obvious that the Simplified Keyboard is easier to master than the Standard Keyboard," comparing QWERTY to an ox and Dvorak to a Jeep.

There's also another matter to consider. During World War Two, August Dvorak was Lieutenant Commander August Dvorak . . . of the United States Navy. Liebowitz and Margolis report that, according to professor Earle Strong of Pennsylvania State University and former chair of the Office Machine Section of the American Standards Association, Dvorak was "the Navy's top expert in the analysis of time and motion studies" and conducted the 1944 studies himself.[7]

Strong conducted his own "carefully controlled"[8] study of the costs and benefits of switching to a Dvorak keyboard in 1956, with the General Services Administration. First, he retrained ten government typists on Dvorak keyboards. While the economics literature on QWERTY versus Dvorak claims the Navy study demonstrated it only took ten days for QWERTY typists to reach their old speed after retraining as Dvorak typists, Strong's group took more than twenty-five days of training at four hours a day to reach their old speeds. Once this group of new Dvorak typists had reached its old speed, Strong started a skill improvement program for both them and a second control group of ten QWERTY typists. Strong found that the Dvorak typists made less progress than the QWERTY typists. He thus concluded that since retraining typists to use Dvorak would never be cost-efficient (though it provides no idea about the cost of training Dvorak typists from scratch), the government should stick with QWERTY. While Liebowitz and Margolis note that this study was pretty much the last nail in Dvorak's coffin regarding the possibility of serious governmental and institutional use of his keyboard, it is not cited in the economics literature on QWERTY versus Dvorak or any of the popular histories.[9]

More recent ergonomics studies, say Liebowitz and Margolis (who cite several examples), "provide evidence that the advantages of the Dvorak is either small or nonexistent." The study they quote at length was conducted by Donald Norman and David Rumelhart, who examined subjects typing on a variety of keyboards, including Dvorak, QWERTY, several alphabetically organized keyboards, and a number of random arrangements. Their study showed that "alphabetically organized keyboards were between 2% and 9% slower than the Sholes keyboard, and the Dvorak keyboard was only about 5% faster than the Sholes," results that are consistent with other recent ergonomics studies. An optimal keyboard, they argue, does three things: it equalizes loads on both hands; it maximizes the amount of work that is done on the "home" (middle) row; and it maximizes the alternation of sequences between hands and minimizes the number of sequences that one finger has to perform. While Dvorak succeeds at the first two requirements (and the figures they provide are that Dvorak typists use their right hand 53 per cent of the time, and that 67 per cent of the time, fingers never have to leave the home row) and QWERTY fails (QWERTY typists use their right hand 43 per cent of the time, and do most of their typing on the top row), Sholes's decision to split up common letter pairs on opposite sides of the keyboard actually makes it highly successful at the third factor, which leads to speedy typing.[10]

Norman and Rumelhart conclude that "for the expert typist, the layout of keys makes surprisingly little difference. There seems no reason to choose Sholes, Dvorak, or alphabetically organized keyboards over one another on the basis of typing speed. It is possible to make a bad keyboard layout, however."[11] Liebowitz and Margolis add that "at the very least, the studies indicate that the speed advantage of Dvorak is not anything like the 20-40 percent that is claimed" and that "the studies suggest that there may be no advantage with the Dvorak keyboard for ordinary typing by skilled typists."[12]

Liebowitz and Margolis close out their already devastatingly thorough reassessment with another argument, that the economists that rely on the example of the QWERTY vs Dvorak debate as an example of the failure of markets to select a "superior" option can only do so because they are relying on "sterile" market models with little or no role for the effects of entrepreneurship, competitive pricing, market research, rentals, mergers, loss-leader pricing, and so on. "In the world created by such a sterile model of competition," they write, "it is not surprising that accidents have considerable permanence." And then, moving in for the kill, they leave a mocking rhetorical question buried in the hearts of David and the other economists they critique: "In such a world, there is ample room for enlightened reasoning, personified by university professors, to improve on the consequences of myriad independent decisions. What credence can possibly be given to a keyboard that has nothing to accredit it but the trials of a group of mechanics and its adoption by millions of typists?"[13]

Regardless of any inefficiencies in the design of his keyboard, by the time Dvorak could bring it to market, the QWERTY keyboard had already become inextricably linked to the operation of the

American economic system. Dvorak's attempt to market a machine capable of reducing both the number of typewriters and typists necessary to run a business during the Great Depression, when typewriter salesmen were buying and destroying used machines in a desperate attempt to generate sales, was not one of the halcyon moments in the history of advertising.

Adding insult to injury, Dvorak's second attempt to market his keyboard in the 1940s also failed, because office workers had become acclimatized to one of the vestigial aspects of the typewriter. In what should have been another instance of brilliant innovation, Dvorak linked his keyboard to Remington's new noiseless typewriter, only to discover that "most typists preferred receiving auditory feedback in the form of the crisp clack associated with the standard typewriter." Shortly afterward (largely due to Earle Strong's recommendations), the U.S. government adopted QWERTY as its standard and ordered 850,000 new typewriters at the outset of World War Two.[14]

Chapter 20

The Poet's Stave and Bar

Like the other technologies of the industrial revolution, typewriting moulds bodies into useful forms in order that they might actually *do* something productive. So it may seem odd, at first, that there are so many associations between the key figures in the history of typewriting and contemporary society's least useful members – poets.

The Wonderful Writing Machine's purple description of Christopher Latham Sholes reads like an attempt to use the word "poet" as many times as possible in one paragraph: "[Sholes] looked more like a poet than any of the things he was or had been. His eyes were sad, like a poet's. He was tall, slender to the point of frailty, with long flowing hair, a short beard, and a medium-length mustache, and he loved poetry although he didn't write it. He also loved puns. His idea of the world's best joke was a poetic pun."[1] Even early typewriter salesmen were apparently worthy of the poet's laurels; the book goes on to describe typewriter salesman C. W. Seamans as "looking like a combination poet and revivalist-meeting preacher."[2]

When Marshall McLuhan writes in 1964 about the boon that typewriting bestowed during "the age of the iron whim," he too turns to the poets for ammunition: "Poets like Charles Olson are eloquent in proclaiming the power of the typewriter to help the poet to indicate exactly the breath, the pauses, the suspension, even, of syllables, the juxtaposition, even, of parts of phrases which he intends, observing that, for the first time, the poet has the stave and the bar that the musician has had."[3]

McLuhan is simultaneously emulating Olson's style and paraphrasing his famous poetic manifesto, "Projective Verse," written in 1959. What McLuhan omits, though, is more telling than what he includes. Olson writes, "It is the advantage of the typewriter that, *due to its rigidity and its space precisions*, it can, for a poet, indicate exactly the breath, the pauses, the suspensions even of syllables, the juxtapositions even of parts of phrases, which he intends" (emphasis added).[4] Here again is the language of discipline: the typewriter enforces rigidity and distribution in space as a means of creating exactitude, of quantizing even empty spaces on a page as a metaphor for breath in a line of oration. And powering and guiding this new regime of control is the "intent" of the poet, herding unruly words into shape. There is little room in such a poetics for the admission of indeterminacy or the role of the reader in the creation of meaning; for Olson, the poet is a master technician in control of every aspect of his or her writing. Not only does a poet "record the listening he has done to his own speech" in a poem, he also indicates, with the help of the preset blanks of the typewritten page, "how he would want any reader, silently or otherwise, to voice his work."[5]

At the same time, Olson instrumentalizes both the body of the poet and the poem that the poet produces into channels for the transmission of information. For Olson, the production of poetry is a matter of utility and relations, and the writer is one component in a larger network shot through with forces and laws:

> It comes to this: the use of a man, by himself and thus by others, lies in how he conceives his relation to nature, that force to which he owes his somewhat small existence . . . [I]f he stays inside himself, if he is contained within his nature as he is a participant in the larger force, he will be able to listen, and his hearing through himself will give him secrets objects share. And by an inverse law his shapes will make their own way.[6]

Olson's language is the language of discipline applied to the task of producing poetry. Through the practice of "Objectism," he plans to dispense with "the lyrical interference of the 'subject' and his soul,"[7] to turn the poet into an efficient channel for the communication of lived experience. The poem is the circuitry that connects the poet-as-recording-device to the reader as receiver: "A poem is energy transferred from where the poet got it . . . by way of the poem itself to, all the way over to, the reader. Okay. Then the poem must, at all points, be a high energy-construct and, at all points, an energy-discharge."[8]

Olson is a pet example for McLuhan because his poetics reinforces one of McLuhan's major contentions, that mid-twentieth-century technologies such as the typewriter, the telephone, the phonograph, and the radio were not merely about extending the control of man, the sovereign subject; they also signalled a "return" to a "post-literate acoustic space."[9] Olson's contention that "if a contemporary poet leaves a space as long as the phrase before it, he means that space to be held, by the breath, an equal length of time" epitomizes this sensibility, the following even more so, as it is mediated by the grid that the typewriter imposes:

> Observe him (i.e., the poet), when he takes advantage of the machine's multiple margins, to juxtapose,

> Sd he:
>   to dream takes no effort
>   to think is easy
>   to act is more difficult
>   but for a man to act after he has taken thought, this!
>   is the most difficult thing of all

> Each of these lines is a progressing of both the meaning and the breathing forward, and then a backing up, without a progress or any kind of movement outside the unit of time local to the idea.[10]

For Olson, what is important about the typewriter is its immediacy. He sees the machine as "the personal and instantaneous recorder of the poet's work,"[11] and a tool with which to restore to both writer and reader the sense of the poet's presence in the finished work, a presence stripped away by the conversion of manuscript to the printed page.[12]

In order to present typewriting in this way, though, Olson has to ignore some explicit evidence in his own examples that typewriting is never about immediacy and breath but always about mediation and writing. It's already implicit in the "invisible" tab stops of the example above, but becomes explicit and visible when Olson writes:

> If [the poet] wishes a pause so light it hardly separates the words, yet does not want a comma – which is an interruption of the meaning rather than the sound of the line – follow him when he uses a symbol the typewriter has ready to hand:
>
> What does not change / is the will to change[13]

The insertion of the virgule (/) is the graphic mark of mechanical mediation in every sense, a solid black bar signifying that there is something in the channel between writer and reader blocking the way, something that both McLuhan and Olson choose to ignore in order to advance an argument for emancipation through rigour.

And what if its insertion was a typo? Even for poets – *especially* for poets – there is always noise in the channel.

One-Finger Typing

The idea of the typewriter as a prosthetic that enables writing is not new to this discussion. It is present from the beginnings of the machine's history as a writing device for the blind, and it persists through McLuhan's notion of technology as "the extensions of man" and in science-fiction scenarios of writing cyborgs. All of these narratives, though, focus on the efficacy of the machine to produce writing. In texts written by the people who actually have little choice but to use the typewriter to communicate, both the writer's mastery over the machine and the ability of the machine to channel the writer's desires are pushed to their limits.

American poet Larry Eigner had cerebral palsy due to a forceps-inflicted injury at birth.[14] His vast oeuvre (more than forty books and hundreds of magazine articles that influenced several major movements in contemporary American poetry) was produced entirely by one-fingered typing with his right index finger. Eigner was strongly influenced by Olson and William Carlos Williams, as he relates in an unpublished letter to Ina Forster:

> Before I read of "energy construct" or maintenance in Charles Olson's "Projective Verse" in the early '50s, in *Poetry New York* (1950), I thought myself that immediacy and force have to take precedence over clarity in a poem (this in reaction to my mother, though I tried or wd've liked to follow, agreed with her insistent advice to be clear), and about the

> same time there was Wm Carlos WIlliams! "A poem is a machine made of words" (he was a medical doctor *ein Arzt?* but he said "machine," not "organism," hm). A piece of language that "works," functions.[15]

To an extent, Eigner agrees with Olson and Williams, viewing the typewriter as a device that preserves a precise record of the poet's thoughts and feelings in a finished poem. Eigner's first encounter in person with Olson was also mediated by typewriter: "I / and my brother visited him once or twice (in '57 or 8 when / I showed him a poem, right away he pulled out his portable / typewriter and copied it!!)"[16] e. e. cummings, Eigner writes, "was really the first to utilize the possibilities for accurate notation – registration – by the typewriter."[17] But there is also a difference in Eigner's poetics born of the fact that writing was not just a metaphorical but a real struggle for him, that truly accurate notation was rarely possible, that writing required real force, real work, to produce a piece of language that worked.

An interesting intersection characterizes Eigner's writing, and the typewriter sits in its middle. On one hand, despite being able to type "fast enough back when to be familiar enough with the keyboard to work in the dark or the dusk with one finger,"[18] the typewriter was barely able to manage the flood of ideas in Eigner's head. "There've always been so many things to do," he writes, noting the reason for his characteristically dense prose, *in* his characteristically dense prose, was that "letters get crowded just from my attempt to save time, i.e., cover less space, avoid putting another sheet in the typewriter for a few more words as I at least hope there will only be."[19] On the other hand, typewriting offers a solution, of sorts, to the problem of its own inability to process the rush of his thoughts: Eigner often resorts to two columns when he writes prose. "It'll be from not deciding or being unable to decide quickly anyway what to say first, or next. Or an afterthought might well be an insert, and thus go in the margin, especially when otherwise you'd need one or more extra words to refer to a topic again."[20] Typewriting may not accommodate *everything* Eigner wants to commit to paper, but the compromise between the two helped to forge a unique poetic style.

Jazz Hands

Poets aren't the only ones enraptured with typewriting in the QWERTY world.

David Sudnow is a sociologist who taught himself jazz piano. In *Talk's Body*, he describes a personal phenomenology of "keyboarding" based on his "daily life on a swiveling chair between two keyboards, that of my piano and that of my typewriter."[21]

Sudnow's guiding trope for both kinds of keyboarding is jazz improvisation. He is not interested in breaking down his music-making movements into therblig-style time and motion study units in order to make keyboarding fit "some existing circuitry model." Instead, he is interested in producing "a new sort of descriptive biology" that might replace the mystifications behind typists' claims to be receiving dictations from ghosts, aliens, muses, giant insects, and other forms of "guidance from above."[22]

There are similarities as well as differences between Sudnow's model and the other perspectives on typewriting examined in previous chapters. Despite the obvious incongruities between a linear keyboard that operates according to a system of major and minor keys, requires the use of foot pedals, and relies heavily on chording effects, and a quadruple-row system of keys of equal value that operate in discrete fashion, Sudnow is determined to demonstrate some congruity between pianos and typewriters. While he's interested in a materialist biological theory of inspiration that would replace the idea of an Outside dictating voice, Sudnow still sees the keyboard as an extension of a sovereign subject, but his model of typewriting is at times closest to the biology of Cronenberg's *Naked Lunch*.

The other major important difference between Sudnow and Olson and the typists that I've already discussed is that the former are *generative* typists. In other words, they compose as they type. They have internalized not only the disciplinary system of touch typing, but also the dictating voice. Whether or not they choose to mystify the dictator and his attendant systems for disciplining the body of the amanuensis into a ready and receptive instrument by presenting that voice as the muse, an alien intelligence, or something else depends entirely on the predilection of the generative typist in question, but it is unquestionably still present. It is in fact the very thing that allows the typist to write, training and informing his or her movements, and always demanding further practice.

Typewriting, for Sudnow, is *embodied* knowledge. He rhapsodizes about "the intelligence of the integrated knowing hand, which guides as it is guided, singing from place to place, making melodies in a network of spatial contexts that are grasped and tacitly appreciated in the most intimate and still mysterious ways."[23] Via a system of touch typing, he trains his body to the point where it responds almost automatically to the keyboard. Once an individual reaches this state, where the disciplinary system has been entirely internalized, he is paradoxically "free" to use both his newly instrumentalized limbs and the typewriter to which they are almost seamlessly joined to compose.

If control begins as a spatial architecture (such as a prison or classroom) designed to transform individuals by progressively objectifying them and subtly partitioning their behaviour on an increasingly fine scale, then that architecture first has to become a set of disciplinary practices that can be internalized into the bodies of the subjects themselves.[24] Even when the architecture itself remains, it is not always necessary. Train a prisoner to believe that he is always being observed, and it is no longer necessary to observe him constantly. Likewise, train a body to type, and it no longer needs to relate to the keyboard as an external architecture. Sudnow types the following:

> When I type the letter "t," my finger does not search for the locale where the "t" is written. Once upon a time, I learned to bring a finger to where "t" is written and then I forgot about its placement in such terms, so much that if I have to fill in a diagram of the typewriter keyboard from imagination today, I must mimic the production of words to rediscover the named keys. When I go for "t" now, I reach, in the course of aiming, toward saying "this" or "that," aiming toward the sounding spot where "t" merely happens to be written. And if I reach for this spot and get somewhere else by mistake, I needn't look at the page to tell. I can feel I have made a wrong reach, just as I can tell I am tripping without having to watch myself.[25]

Sudnow has interpellated the machine's disciplinary systems so thoroughly that he is loath not only to describe his "hand's knowledge" in terms of the "topography of the keyboard," he won't even use the language of music to describe the spatializations his body has imagined as musical "notes" because "these [terms] divide the keyboard from the body."[26]

Sudnow's sense of what he is doing is not all that different from Olson's Objectism; both assume a more or less unproblematic transmission of direct experience as content from writer to reader. Sudnow claims that if you "use the touch-typing method to copy over the sentence you are now reading . . . you get almost as close to a recoverable notation system as you can get." Sudnow does note that typescript provides no clues as to the temporality of the writing; "I can produce the sentences you are now reading in fifteen seconds or fifteen minutes, produce the first portion in a rapid fire and the rest after a coffee break, and you cannot tell." For him, handwriting still is the privileged sign that indicates the passage of time. In order to gain some knowledge of the "temporal structure" of a piece of writing, Sudnow claims "You come still closer if you try to reproduce sights such 'as these' [the words are in cursive text] in their particularity."[27] However, Sudnow does not account for the very distancing factor that Olson bemoans: mass print publication, which turns "these words" into yet another infinitely reproducible sign, even if it is a sign that evokes handwriting. Jacques Derrida famously makes this same point at the end of his essay "Signature Event Context" by reproducing his own signature to demonstrate that the very things that Olson and Sudnow champion – the effects of performance, presence, and speech on a text – presuppose the very things they hope to exclude: error, slippage, reproducibility, and multiplicity – the effects of writing, typewriting included.[28]

Regardless of the flaws in Sudnow's argument about what is transpiring, the statements that his text makes are still fascinating. All of the discipline he exercises on himself goes to an interesting end: the aesthetically pleasing but decidedly non-utilitarian creation of a body capable of making art: "[W]hen fingers in particular learn piano spaces in particular, much more is in fact being learned about than fingers, this keyboard, these sizes. A music-making body is being fashioned."[29] Note that as Sudnow accedes to the discipline of his two keyboards, his grammar also becomes passive, and his descriptions of his own body become increasingly objective:

> [M]y articulating organs are now set up in a precise spatial scaling. The finger feels the width of a key at the piano, perhaps assessing the key's extent by feeling the edge of the next key. The hand is now toned up for such sizes all through the domain. The depths and textures of the places are known. The hand accordingly assumes a sort of roundness and balance appropriate for speaking.[30]

Once Sudnow has explained his process, he performs it for the reader. The entirety of chapter 36 of *Talk's Body* is an improvisational performance, a jazz for the typewriter, documented with a video camera (or so the text tells us – shades of Gilbreth here). The act of typewriting itself becomes a performance. That written performance record is replete with signs of discipline attempting to steer a wayward body toward a desired end. Sudnow has left in all typos, included spaces, and has left the right-hand margin ragged; he asks that the reader "treat the errors here as a signt t t signt that sa a struggle is taking place."[31]

However, these signs *remain* as signs of a struggle that perhaps took place elsewhere, in another medium, if it indeed took place at all. Even assuming that this text is not a simulation, the physical qualities of a typescript are very different from those of the printed book, and translating the former into the latter inevitably creates all sorts of slippages and gaps. Despite the aforementioned attempts to make chapter 36 evoke a typescript, it bears the unmistakable signs of typesetting, including, most tellingly, ligatures that link multiple letters together into one character (for example, ff, fi, fl, ffi, ffl).

Like many typists, Sudnow struggles to use the rigour of typewriting to produce art, in the hopes of producing truth through art's beauty. But, to return to an earlier theme, typing has a problematic relationship to truth, even though we often assume that it will produce it for us. It's time to take a closer look at that problem.

Chapter 21

Typewriting, Identity, and Truth

What underlies this attempt to demonstrate the importance of typewriting is an assumption that typewriting produces truth, that quality that the Western concept of language has always associated with the things Sudnow and Olson champion as the hallmarks of their writing: presence, breath, speech.[1] This is an argument that Jacques Derrida has thoroughly dismantled, demonstrating that these privileged terms are founded on the indeterminacies and slippages of writing. Derrida's arguments, however, did not keep entire generations of writers from portraying the typewriter as a device that extracts truth from even the most reluctant of bodies, unceasingly *pulling* it out, as Henry James might have said.

Cases of Identity

"It is a curious thing," says Sherlock Holmes in "A Case of Identity," his third adventure, "that a typewriter has really quite as much individuality as a man's handwriting. Unless they are quite new, no two of them write exactly alike. Some letters get more worn than others, and some wear only on one side."[2] Conan Doyle wrote this statement in 1891, only a few years after the first commercial typewriters found their way onto the market.

"Truth," says Michel Foucault, "is a thing of this world." Each society has its own regimes of truth, its own mechanisms for distinguishing true statements from false statements, its own ways of distinguishing who is allowed to pronounce truth. Truth, in other words, is not something external to events, not something that obtains in all times and places, but something that each society *produces* for itself, through the application of mechanisms such as exemplary material evidence, expert testimony, and judgment in the courtroom.[3] A full two years after Conan Doyle writes that by studying the idiosyncrasies of typewritten text under a magnifying lens, it is possible to identify the typewriter that produced a given document, a court case in New Jersey, *Levy v. Rust*, produced the same conclusion. In the words of the judge,

> An expert in typewriting is brought here and that expert sat down by my side at the table here and explained his criticisms on this typewriting, and I went over it with him carefully with the glass . . . it appeared very clearly. I was very much struck by his evidence. If you compare the typewriting work, it contains precisely the same peculiarities which are found in the typewriting in these seven suspected papers.[4]

Though the system for identifying the unique characteristics of a typewriter is similar to the one used for fingerprints – a comparison of patterns and irregularities in the prints or impressions that fingers and type slugs, respectively, leave – the first criminal trial that resulted in a conviction based primarily on *fingerprint* evidence did not occur until 1901, *eight years after* the *Levy v. Rust* case.[5] In the eyes of the law, the impressions left by writing machines were recognized as unique before those made by the people that use them.

Fingerprinting was originally not a criminal technology but an instrument of colonialism, and a technology of surveillance rather than of investigation. In 1858, a British magistrate named William Herschel stationed in a Bengal village had noticed that the Bengalis had a tradition of using one fingerprint as a signature. Reasoning that each fingerprint must be unique, and anticipating difficulties with a particular individual, he ordered a complete set of that person's finger and palm prints.[6]

The notion that fingerprints are unique and admissible as evidence is based on a mathematical study conducted around 1890 by Francis Galton, a cousin of Charles Darwin. The following year, Edward Henry, a Bengali police officer who later became assistant commissioner of the London Metropolitan police, expanded Herschel's work into a system of classification. By 1900, Scotland Yard was using what eventually was dubbed the Galton-Henry system, later refined into the FBI-NCIC (National Crime Information Center) fingerprinting system.[7]

A decade later in the United States, The 62nd Congress enacted the United States Statute of 1913, Chapter 79, which permitted "the introduction of admitted or proven handwriting exemplars for comparative purposes." By court decisions, this statute was extended to cover typewriting. In *People v. Werblow* (1925), it was stated that "the law is well settled that such specimens of typewriting are properly received in evidence for the purposes of comparison." The first known attempt to commit a forgery with a typewriter occurred only two years later. A New York lawyer named Risley used his Underwood typewriter to add the words "the same" (an interesting fact in and of itself) to an affidavit written by another lawyer who used a Remington typewriter. An expert witness named William J. Kinsley described thirteen respects in which the type produced by the machines differed, and further identified Risley's typewriter as the machine that had added the additional two words.[8]

Before the trial, Risley had approached Arthur W. Blackwell of the Blackwell Typewriter Company and asked him to attempt to produce a machine that could duplicate the text produced by his own Underwood as an attempt to demonstrate that all typewriters were *not* unique. Blackwell complied, but the court expert was able to identify how the machine differed from Risley's original. Even if Blackwell had succeeded, though, it's uncertain the outcome would have been any different, as a much more infamous case in American history demonstrates.[9]

Woodstock No. 230,099

Alger Hiss was one of the most important diplomats of the New Deal era. A lawyer by training, he played many important roles in the Roosevelt administration's post-war reconstruction plans, ranging from investigating profiteering by the munitions industry to establishing the United Nations in his role as deputy director of the State Department's Office of Special Political Affairs. In 1946, Hiss left government service to become the president of the Carnegie Endowment for International Peace.

In 1948, Whittaker Chambers testified before Senator Joe McCarthy's House Un-American Activities Committee (HUAC) that Hiss was a communist and a spy. Hiss not only denied the charge, he sued Chambers for libel. Chambers then produced copies of a series of State Department documents, which he claimed Hiss gave him with instructions to send them to the Soviet Union. When Hiss appeared before a grand jury to deny committing espionage, he was charged with perjury. His first trial ended in a hung jury, but he was retried, found guilty on two counts of perjury, disbarred, and sentenced to forty-four months in jail.[10]

The star witness against Hiss was Woodstock typewriter model No. 230,099, purportedly the typewriter that the Hiss family had owned at the time the incriminated documents were supposed to have been produced. The typewriter, however, had been out of Hiss's possession for many years; Hiss had given it to Perry and Raymond Catlett, the mechanically-inclined sons of their maid, in December of 1937. In 1945, a trucker named Ira Lockey found it in a back yard in the rain and took it as partial payment for a moving job. Six weeks before the trial began, Hiss's lawyers found it at Lockey's house. Ironically, though the defense produced the typewriter, it became a major exhibit for the prosecution. Of the forty-three documents that Chambers claimed Hiss gave him to send the Soviets, the prosecution claimed that forty-two of them had been typed on this machine by Priscilla Hiss, Alger's wife. The machine sat on the trial table throughout the extremely lengthy proceedings (more than 2,300,000 words of testimony). In his closing remarks to the jury, prosecutor Thomas F. Murphy pointed dramatically at the machine and told the jury that the Woodstock was "the immutable witness forever against" Hiss. Before sentencing, Hiss made one simple statement: "I am confident that in the future all the facts will be brought out to show how Whittaker Chambers was able to commit forgery by typewriter. Thank you, sir."[11]

Enter Martin Tytell, of Tytell's typewriter shop in New York. Tytell had filled a wide number of bizarre requests for typewriter modifications over the years, from musicians who wanted keyboards with musical notes to a mystery writer who wanted a keyboard that consisted of nothing but crosses and bones, to a man who asked him to build a typewriter with a question mark on every key.[12]

In March of 1950, about two months after Hiss's conviction, Mr. Chester T. Lane, Hiss's defence lawyer, approached Tytell to ask if he could create a machine that would exactly duplicate the type produced by Hiss's machine, working solely from the samples produced by that machine. Tytell agreed, and for a fee of $7,500 he set to work.

He began with a Woodstock model with a serial number close to Hiss's machine drawn from his own "typewriter morgue" and enlisted the help of an engraver to copy all of the type-face defects

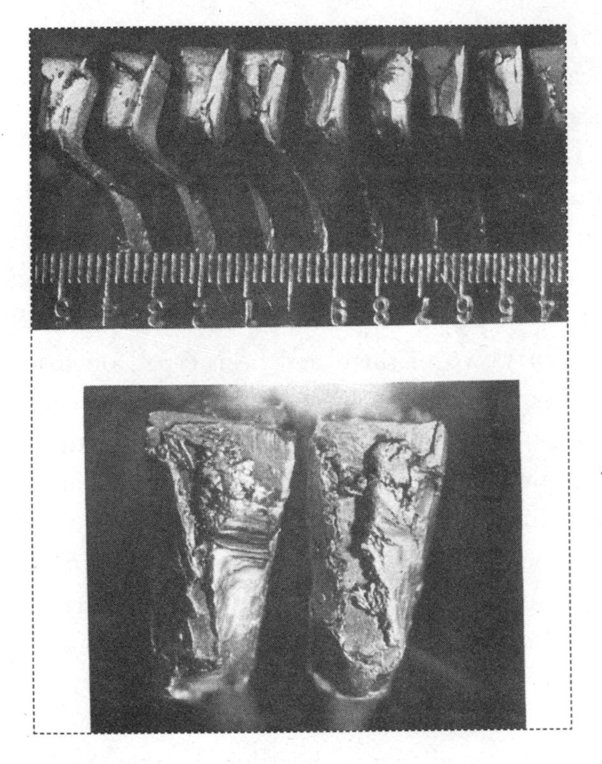

The Woodstock machine that was presented as Hiss's typewriter during his trials exhibited unusual soldering on the keys. Hiss believed that it had been deliberately altered as part of the effort to incriminate him.

and characteristics of the Hiss machine onto the type faces from his scavenged Woodstock. He subsequently soldered the forged type faces back on to the type bars of the scavenged Woodstock and begin the painstaking process of adjusting the machine's movements to match the Hiss machine.

During the adjustment process, Tytell would occasionally take his work in progress to test against the opinions of one of several prestigious document examiners. He found, to his surprise, that "all document examiners I had visited refused a professional assignment to assist me." Instead, they berated him and accused him of "dangerous" and "illegal" activities. It was only then that it began to occur to Tytell how strong the anti-Communist sentiments and paranoia in the United States actually were.[13] Things became worse: Tytell's engraver caught tuberculosis and, when no other engraver would take the job, he was forced to do it himself, adding at least a year to the project as he sorted through and modified forty-two pieces of type from over two thousand that he had collected.

Tytell's adjustments became more and more minute. He adjusted the type bars to within a thousandth of an inch of the measurements from Hiss's machine. He borrowed a tool called the Ames Densimeter, one of perhaps twelve in existence at the time, to determine the precise density of the rubber platen roller so as to match the roller on Hiss's machine, which would create a "microscopic" difference in the vertical spacing of typewritten lines on the page as it gripped the page when advancing. He adjusted the "escapement" (the combination of parts beneath and behind a typewriter that control its inter-letter spacing) to match the tendency of Hiss's machine to "creep" or crowd letters together as it moved closer to the right-hand margin.[14]

Tytell finally completed his machine, produced a set of specimens "designed to put document experts to the supreme scientific test," then deposited the machine in a Marine Midland Bank vault. One expert, Mrs. Evelyn S. Ehrlich, Harvard University's Fogg Museum of Art's forgery expert, testified that "except for subtle details, I found that microscopic variations on one machine had been duplicated on the other so faithfully that I might not have believed it possible if I had not been informed that two machines were involved."[15]

The final expert Mr. Lane consulted was Dr. Daniel Norman, the Director of Chemical Research of the New England Spectrochemical Laboratories. Dr. Norman's examination of Woodstock No. 230,099 led Lane to a conclusion he had been suspecting for some time: that the machine itself was a forgery, created to forge the typewritten documents that were being used to incriminate Hiss.

Dr. Normal wrote in his affidavit that this typewriter was "not a machine which has worn normally since leaving the factory, but shows positive signs of having been deliberately altered, in that many of its types are replacements of the originals and have been deliberately shaped."[16] As on most typewriters, the type on Woodstock machines consisted of a smalls slug that was soldered into place on the end of each type bar. On No. 230,099, most of the type slugs had "been soldered onto the typebars in a careless fashion, quite unlike the kind of soldering job done at the Woodstock factory or in regular repair operation." Dr. Norman was also able to determine that the solder on the altered slugs had a different metallic content than the solder on the apparently unaltered ones, and from that used on contemporary machines produced by Woodstock. The metal in the type slugs themselves also contained elements that did not appear in Woodstock-manufactured machines until much later than the machines manufactured around the time of No. 230,099. Finally, Dr. Norman also found tool marks on the striking faces of the letters and unusual finishing or polishing uncharacteristic of machines that had aged in normal fashion.[17] In sum, the evidence would have been enough for Grissom on *CSI*.

Mr. Lane speculated in his affidavit that the most likely time for the machine to have been altered was *before* the machine had been located by the defence, probably between Chambers's first testimony before HUAC in August 1948 and the day that he produced the incriminating documents in November of the same year. "The fact that between November and April neither the defense nor the thirty-five agents of the FBI could fine the machine," he writes, "suggest that it was during this period that further work was being done on the types, in an effort to remove at least the more obvious tool marks which would betray the deception."[18] The judge, however, remained unimpressed.

After Hiss's conviction, his lawyer filed unsuccessfully for a retrial. Between his conviction and the mid-80s, Hiss was refused a retrial by the higher courts on eight occasions. The disagreement came down to a war of conflicting authoritative statements. In 1976

Richard Nixon said, "A typewriter is . . . almost the same as a fingerprint. It is impossible, according to experts in the field, to duplicate exactly the characteristics of one typewriter by manufacturing another one."[19] Nixon's veracity, of course, has been questioned before. There is at least one other factor worth considering: a 1960 memorandum from J. Edgar Hoover revealing that the FBI had the capability to commit typewriter forgeries since at least 1959.[20]

Though Hiss was readmitted to the Massachusetts bar in 1975, he died in 1996 without having cleared his name.

The reality of the relationship between typewriting and truth is that typewriting, like anything else that can be read, can also be forged. But that's not how we mythologize it. For the two-hundred-odd years that typewriting *was* writing, the texts that we produced nevertheless attempt to convince us that the pull of typewriting is strong enough that it can even extract language – and possibly truth – from animals.

Chapter 22

The Monkeys

Sooner or later, anyone writing about typewriting has to deal with the monkeys.

In 1909, French mathematician Emile Borel published a book on probability. As an illustration of a special case of a mathematical proposition named "Kolmogorov's zero-one law," Borel invented the image of the "dactylographic monkey" ("dactylographie" being French for "typing"). He then stated that, according to the zero-one law, the probability that this typing monkey would eventually reproduce every book in France's National Library was equal to one.[1]

Ever since, the image of the typing monkey has obsessed the popular imagination. The number of monkeys and typewriters has proliferated, though in order for the theorem to be true, there doesn't have to be an infinite number of typing monkeys, only one monkey who makes an infinite number of keystrokes. The location and size of the monkeys' oeuvre has also varied. In some accounts, the monkeys are retyping the contents of the Library of Congress, or, as in Russell Maloney's short story "Inflexible Logic,"[2] the British Museum.

In Maloney's tale, written in 1940, Mr. Bainbridge, a thirty-eight-year-old independently wealthy gentleman of leisure, overhears at a

New York cocktail party the voice of Bernard Weiss, a critic, saying, "Of course he wrote one good novel. It's not surprising. After all, we know that if six chimpanzees were set to work pounding six typewriters at random, they would, in a million years, write all the books in the British Museum." Interestingly, Bainbridge immediately mishears the already incorrect exemplum: "What's this about a million chimpanzees, Weiss?" The story is clarified for him, and, being both a man who "read quite a bit of popular science" and also a man of means, Bainbridge immediately procures a circus trainer and six chimps, sets up a dormitory for them behind his stable, and clears a workspace in his conservatory, complete with chairs of "the unpadded, spring-backed kind favored by experienced stenographers" and all the other accoutrements of the modern office, including a water cooler, a rack of paper cups, and (as a concession to the taste of the employees) bunches of ripe bananas.

Several weeks later, he invites his friend James Mallard, an assistant professor of mathematics at New Haven, to observe. Bainbridge shows Mallard that the chimps have produced, in less than three

The archetypal typing monkey.

months, "The prose of John Donne, some Anatole France, Conan Doyle, Galen, the collected plays of Somerset Maugham, Marcel Proust, the memoirs of the late Marie of Rumania, and a monograph by a Dr. Wiley on the marsh grasses of Maine and Massachusetts," *without spoiling a single sheet of paper*. This is the salient point of this story: the presumed ability of typewriting to elicit, from chimpanzees, flawlessly reproduced ("true") works of literature, philosophy, science, and natural history, without so much as a wasted sheet of paper or even a single line of typographic nonsense. Moreover, the editions that the chimps produce are definitive: "He was leafing through a manuscript that had been completed the previous day by Chimpanzee D, Corky. It was the complete diary of Samuel Pepys, and Mr. Bainbridge was chuckling over the naughty passages, which were omitted in his own edition."

The flawlessness of the chimps' output begins to wear on Mallard, who predicts, "These chimpanzees will begin to compose gibberish quite soon. It is bound to happen. Science tells us so" (the last bit makes him sound like Frank Gilbreth explaining touch typing to his children; the notion that typewriting produces miraculous results is inevitably bolstered by the global invocation of "science"). When the chimps continue to produce flawless manuscripts, Mallard becomes completely unhinged and barges into the conservatory armed with a pair of .38 calibre revolvers. "It is certainly almost beyond the bounds of credibility that these chimpanzees should write books without a single error, but that abnormality may be corrected by – these!" He slays Bainbridge and all of the monkeys, but not before Bainbridge fatally wounds him as well. In the last lines of the story, the last survivor, Chimpanzee F, with his final erg of energy, begins to type, with one finger, the narrative analogy for a desire for its own emancipation: "UNCLE TOM'S CABIN, by Harriet Beecher Stowe. Chapte . . . Then he, too, was dead." (There is no mention of whether or not the revolvers and the typewriters in question were both Remingtons. Remington had been manufacturing .38s since 1878,[3] the same year "that the typewriter passed the experimental point and became practical."[4] That Mallard chooses to correct the output of the one with the output of the other confirms the familiarity between the two devices.)

The Blurst of Times

There are few indications of a statement's degree of importance to contemporary popular culture than its relationship to *The Simpsons*. People repeat snippets of *Simpsons* dialogue to each other like the commonplaces and aphorisms of old; sometimes it seems that all the world is Springfield. It therefore warms the cockles of my seedy little heart that Episode 76, "Last Exit to Springfield," one of the most popular episodes in the history of one of the most popular of all television shows, contains a typing monkey joke. While showing Homer around his vast mansion, as if to demonstrate its near-infinite capacity, and, by extension, the size of his personal fortune, the despotic Mr. Burns does a Bainbridge and opens a door onto a room full of typing monkeys. A monkey hands a piece of paper to him, and Burns reads aloud from it: " 'It was the best of times, it was the *blurst* of times?!' You stupid monkey!"[5]

In most recent popular versions of Borel's tableau, the monkeys are retyping the complete works of William Shakespeare.[6] One of the most famous of these references is the last line of chapter 9 of Douglas Adams's *The Hitch Hiker's Guide to the Galaxy*. The protagonists, Arthur Dent and Ford Prefect, have just been rescued from certain death in the void of intergalactic space by a ship powered by an Infinite Improbability Drive. As the ship reaches a speed of "two to the power of twenty thousand to one against and falling," the following highly (but not infinitely) improbable event occurs:

> Arthur had jammed himself against the door to the cubicle, trying to hold it closed, but it was ill fitting. Tiny furry little

> hands were squeezing themselves through the cracks, their fingers were inkstained; tiny voices chattered insanely.
>
> Arthur looked up.
>
> "Ford!" he said, "there's an infinite number of monkeys outside who want to talk to us about this script for *Hamlet* they've worked out."[7]

Perhaps in response to Adams, Italian-American mathematician and philosopher Gian-Carlo Rota had written in a manuscript that was unpublished at the time of his death in 1999 that "if the monkey could type one keystroke every nanosecond, the expected waiting time until the monkey types out *Hamlet* is so long that the estimated age of the universe is insignificant by comparison . . . this is not a practical method for writing plays."[8]

But Rota wasn't the first mathematician to dally with the typing monkeys, and he probably won't be the last. In 1979, Dr. William R. Bennett, Jr., a physics professor and computer scientist at Yale, calculated that "if a trillion monkeys were to type ten randomly chosen characters a second it would take, on the average, more than a trillion times as long as the universe has been in existence just to produce the sentence 'To be or not to be, that is the question.'"[9] W. J. ReMine concurs; he writes that taking only the lower-case letters and spaces into account, the probability of typing even the first hundred characters of *Hamlet* "is one chance in 27^{100} . . . If each proton in the observable universe were a typing monkey (roughly 10^{80} in all), and they typed 500 characters per minute (faster than the fastest secretary), around the clock for 20 billion years, then all the monkeys together could make $5 \cdot 10^{96}$ attempts at the 100 characters. It would require an additional $3 \cdot 10^{46}$ such universes to have an even chance at success."[10]

Poet, programmer, and pataphysician Neil Hennessy goes a step further, fusing the typing monkey problem onto another phenomenon I'll describe in the next chapter: the typists' duel.

Hennessy begins with the cheeky observation that, despite the aforementioned statistical unlikelihood, "in 1602, Willy the Typing Monkey entered *Hamlet* into the Stationer's Register of England, much to the chagrin of the calculators of probability." Given the fact of the play's existence, he decides to focus on another issue: how to measure Willy's achievement.

Sticking to the first hundred characters of the play, Hennessy's first observation is that, probabilistically speaking, if Willy typed "the tragedie of hamlet actus primus scoena prima enter barnardo and francisco two centinels barnardo," it is equally likely his associate would type "abababababababababababababababababab." The question Hennessy raises is, should the experimenters reward the monkeys equally for the texts they have produced? If Willy realizes that he's going to receive the same number of bananas for meaningless strings of letters as for major tragedies, he'll be less inclined to continue to type interesting literature for us.

This raises a larger question: how does one assess the relative merits of the texts produced by the monkeys in the typing pool, in terms of their degree of complexity? Hennessy proceeds to compare Willy the Monkey's *Hamlet* to Jimmy the Monkey's *Finnegans Wake*. Returning to the theories of Kolmogorov (the mathematician whose work the typing monkey problem was formulated to explain in the first place), Hennessy notes that the Kolmogorov complexity (K) of an object *x* is the length of the shortest binary program that outputs that object; however, because Kolmogorov proved in his Noncomputability Theorem that K is uncomputable, there must be an analogous measurement that we *can* actually calculate. Since Kolmogorov proved that stochastic entropy and expected algorithmic complexity are equal, Hennessy substitutes Shannon's stochastic entropy (H) for K to produce a way to approximate the K value for *Hamlet* and FW^1 (a set of the first 138,902 characters of *Finnegans Wake*, which is the same length as *Hamlet*).

According to Hennessy's calculations, on average, a text that contains all the words in *Hamlet* will be less complex than a text that contains all the words in *FW*[1] – not surprising, because there are about half as many unique words in *Hamlet*. Hennessy, however, is not satisfied: "It would be premature to conclude that *Hamlet* is less complex than *Finnegans Wake*, because the stochastic entropy of the words does not take into account their syntactic relations with each other." He then searches for a method of comparing the entropy of entire sentences and, after dabbling with Natural Language Processing n-gram models, he settles on Kontoyiannis's method for entropy estimation of an entire text via string matching. After a lot of math that's well beyond my ability to comprehend, Hennessy awards the Top Banana prize to Jimmy the Monkey for producing the work with greater complexity.

If you're wondering what you might see if you looked over Willy the Monkey's shoulder as he types on tirelessly into the infinite future, point your web browser to The Monkey Shakespeare Simulator,[11] which began operation on July 1, 2003, with one hundred simulated monkey employees. In the simulator, time passes 86,400 times faster than it does in the physical world. The assumption that the programmers made is that each monkey types at an average but respectable rate of 60 words per minute, and that each page requires two thousand keystrokes to fill. In between typing, the monkeys, being monkeys, procreate, so the typing pool is perpetually growing. The longer any individual runs the simulator, the better the odds of successfully producing a fragment of Shakespeare. As of this writing, the record for the longest meaningful Shakespearean string belongs to the virtual monkeys overseen by Dan Oliver of Scottsdale, Arizona. On August 4, 2004, Oliver's monkeys produced 19 letters from *Two Gentlemen of Verona* after 42,162,500,000 billion billion monkey-years: "VALENTINE. Cease to1dor:eFLPoFRjWK78aXzVOwm)-';8.t . . ." matched "VALENTINE. Cease to persuade, my loving Proteus: Home-keeping youth have ever homely wits. Were't not affection chains thy tender days To the sweet glances of thy honour'd love . . ." Other contenders include 18 letters from *Timon of Athens*, 17 letters from *Troilus and Cressida* and 16 letters from *King Richard II*.[12]

Outside of the realms of mathematical theory and computer simulation, a group of scientists at Plymouth University (U.K.)'s MediaLab, in conjunction with the Paignton Zoo in Devon, brought life to Bainbridge's fictional experiment in 2003, albeit in less posh surroundings. For four weeks, six Sulawesi crested macaque monkeys – Elmo, Gum, Heather, Holly, Mistletoe, and Rowan – had at their disposal one computer, which hardly seems like adequate technical support, given the ambitious aims of the project. (The experiment provides no word on whether it was a Mac or PC, what type of software the monkeys could access, and so on.) Vicki Melfi, a biologist at the zoo, pointed out that macaques are an ideal choice for such an experiment, because "they are very intentional, deliberate and very dexterous, so they do want to interact with stuff you give them."[13] During the experiment, live updates of the macaques' work were published on the web, together with a live webcam view of the production scene.[14]

The macaques produced five pages of text between them, primarily filled with the letter S (many of the monkeys chose to sit on the keyboard rather than press the keys in a conventional manner, which may have resulted in long repetitive strings of single characters). However, "There were greater signs of creativity towards the end, with the letters A, J, L and M making fleeting appearances." Geoff Cox, the designer of the test, said that it wasn't actually an experiment "as such" but a performance designed to demonstrate the difference between animals and machines. "The monkeys aren't reducible to a random process," he said. "They get bored and they shit on the keyboard rather than type."[15]

Despite falling "victim to the distractions which plague many budding novelists" – climbing frames, ropes, and toys that provide more satisfying interactivity than a word processor – the fruits of the macaques' labour was actually published as a chapbook titled *Notes Towards the Complete Works of Shakespeare* (Copyright © the authors, 2002, ISBN 0-9541181-2-X). The complete text is also available online.[16]

Granted, there are problems with this scenario. For human, chimp, or macaque, typing on a word processor is very different from typing on a typewriter, so the results of this experiment may not be all that relevant. The underlying assumptions and statements by the formulators of the infinite monkey problem and the macaque experiment are, as usual, where the real substance lies.

Animal rights activism has been a strong element of contemporary British culture for many years (too bad they had to kill off nearly all their own native species and a good number of those in their colonies before they became concerned about world wildlife in general, but that's the way it goes). The notes to *Notes Towards the Complete Works of Shakespeare* suggest that the ethos of animal rights education is actually what underlies the macaque "experiment": "though it appears to test the truth of the formula, in reality it emphasizes the unreliability of human (scientific) hypotheses. Animals are not simply metaphors for human endeavour. The joke (if indeed there is one) must not be seen to be at the expense of the monkeys but on the popular interest in the idea – especially those in the computer science and mathematics community (interested in chance, randomness, autonomous systems, and artificial life)."[17] Monkeys have their own subjectivity, then, and, despite the fact that they would just as soon type with their asses, do not deserve to be objectified or turned into the butt of our jokes. The implication is that *we* are the fools for aping scientific principles without fully understanding the implications.

As do many texts produced through and around typewriting, the notes to the *Notes Towards the Complete Works of Shakespeare* experiment also raise questions about canonical literary authority and authorship. "William Shakespeare" is one of the most successful Western author-*functions* of all time, and *Hamlet* the jewel in its crown, but people inevitably conflate that function (the name under which critics organize a series of texts) with the writer (at the close of the year 2000, *Time* famously named William Shakespeare the person of the millennium). The experimenters make the following assertions:

> The fact that the work of Shakespeare is probably the work of a group of writers working under a pseudonym adds further irony to the work. Clearly, Shakespeare did not produce his works by some chance operation but it is also entirely disputable that he existed at all or certainly that he was one person in fact, one commonly held view is that he was an illiterate actor and a consortium of writers used his name as an ironic joke. The creative thinking subject as the site of consciousness, and the subject as a crucial part of a sentence and text – that which the action is determined by – remains a contested and contradictory set of ideas. Creativity is neither random nor entirely predetermined, in other words.[18]

While the claim for factual status and the invocation of clarity may be overstating their case, the experimenters do make some valid points. Authorship is a complex phenomenon, and the ongoing validity of the sovereign creative subject is very much a contested notion. However, it may also be time to examine the ongoing attempt to associate typewriting with personal dignity.

We've Got Something to Say

The final infinite-monkey-related text that I want to discuss, by way of summary, is David Ives's very funny and slyly intelligent one-act play, *Words, Words, Words*.[19] If The Monkey Shakespeare Simulator provides some idea of what the dactylographic monkeys' typescripts might look like, Ives provides some insights into the conversations that they might have with each other while working. In the process, he raises virtually all of the issues that arise in other discussions about typing monkeys and provides some insights along the way.

The play opens on three typing monkeys: Milton, Swift, and Kafka (a female monkey). The appearance of all three is both anthropomorphized and infantilized; instead of dressing in monkey suits the actors portraying them "wear the sort of little-kid clothes that chimps wear in circuses."[20] Each types at a separate speed for a few minutes, then they begin to speak.

The first exchange of interest concerns the question of what constitutes sense and nonsense in a typewritten milieu. Swift reads, "'Ping drobba fft fft fft inglewarp carcinoma.' That's as far as I got"; Kafka responds, "I like the 'fft fft fft,'" and Milton concurs: "Yeah. Kind of onomatopoeic."[21] These are evidently monkeys versed in the full range of poetic texts produced with the aid of typewriting (the most radical of which I'll discuss in the next chapter). They realize, from the outset, that there are many strategies a reader can use to derive meaning from a poetic text that have nothing to do with normative syntactic readings. Onomatopoeia is as legitimate a criteria for reading as any other. The sequence "Ping drobba fft fft fft," if treated as a sound poem, could easily be interpreted as a mimicry of the operation of the typewriter itself, from the bell at the end of the line ("ping") to the carriage return ("drobba") to the sound of tabbing across the page or the advancement of paper around the platen ("fft fft fft"). Even "inglewarp carcinoma" has an internal "ar" rhyme to recommend it as a sonic combination. As a word pair, it would hardly be out of place in many volumes of innovative poetry written over the last thirty years.

When Kafka reads her own composition – "K.K.K.K.K.K.K.K.K.-K.K.K.K.K.K." (which, in the context of Franz Kafka's writing, I read not as nonsense, but as the repeated typed signature of Joseph

K., which performs some of Kafka's major themes – the agonism between individual identity and bureaucratic anonymity; the transformation of the beleaguered individual [K.] into an instrument of terror [K.K.K.] through fascism) – Swift responds, "What is that – postmodernism?" Not even; it's high modernism, but this is the first monkey-related text that has bothered to question the terms of the argument itself, namely, the status of the sensible. Swift observes to Kafka, "At least it'll fuck up his [Dr. Rosenbaum, the experimenter's] data."[22]

The chief concern of this trio of monkeys is epistemological in nature: "What is *Hamlet*?"[23] All they know is that it is something they are required to produce by typing, even though they also know that none of them would recognize it if they did produce it, and they hope that its production is also the condition of their eventual freedom from servitude and observation. When the monkeys begin to speculate on the conditions of their release if they *do* produce *Hamlet*, Swift worries (in anticipation of Hennessy), "Or will they move us on to *Ulysses*? (*They shriek in terror at the thought*)."[24] If the popular imagination holds *Hamlet* as the apex of Western literary creation, then Joyce (even the relatively accessible *Ulysses* rather than the *Wake*) remains, sadly, the popular example of a limit case for legibility.

For these monkeys, the role of amanuensis or word-processing clerk is demeaning, because it belies the possibility of individual creativity. This is Swift's primary complaint. "Does Dr. Rosenbaum care about voice? Does he care about anybody's individual creativity? . . . No! He brings us in here to produce copy, then all he wants is a clean draft of somebody else's stuff."[25] Swift, like the sixties proto–boy band The Monkees, asserts that he has "something to say," but never gets around to specifying on paper exactly what that *something* is. Part of the humour of this piece depends, as anthropomorphic comedy often does, on the fact that the monkeys remain monkeys, despite their aspirations. In a parody of behaviourist experiments, the monkeys are rewarded for the occasional antic display. But like prisoners in a penitentiary, their desired currency is cigarettes. Swift, ever the pragmatist despite his revolutionary rhetoric, observes, "You do your Bonzo routine and get a Gauloise out of it. Last week I totalled a typewriter and got a whole carton of Marlboros." As with *Royal Road Test*, even the destruction of the tools that the monkeys write with becomes a useful part of the discourse for the scientists observing them. When Milton cracks, "The trouble was, you didn't smoke 'em, you took a crap on 'em" (like the macaques), Swift replies, "It was a political statement."[26] Over the axis of typewriting, the play seesaws back and forth over some big questions. What is human and what is animal? What is sensible and what is nonsensical? What is revolutionary and what is reactionary?

The genius of the play is this: by being placed in a *Hamlet*-like situation – a harsh and oppressive world ruled by intricate webs of political power – the three monkeys begin, though the same process of posing rhetorical questions that serves as the basis for Hamlet's soliloquies, to *speak* sections of the text of *Hamlet* unwittingly as they describe their situation to each other. As in Objectist poetry and typewritten dictation, in *Words, Words, Words*, speech rather than writing produces truth. After the *Ulysses* joke, Swift proceeds straight into a pastiche of the Hecuba speech: "What's *Hamlet* to them or they to *Hamlet* that we should care? Boy, there's the respect that makes calamity of so long life! For who would bear the whips and scorns of time, the oppressor's wrong, the proud man's contumely –"[27] As in *Hamlet* and its famous postmodern intertext, *Rosencrantz and Guildenstern Are Dead*, there are plays within plays here, and the audience delights at the nested layers of dramatic irony. Their epistemology produces truth, and yet, as the monkeys also accurately predicted, when they produce *Hamlet*, they are unable to recognize it.

Further, they not only speak portions of the play, they also begin to enact it by plotting "*revenge*" on Rosenbaum, whose name evokes

"Rosencrantz" and whose role in the chimps' lives evokes Claudius. While there is no Guildenstern in sight, Rosenbaum may soon be dead. Swift plans to poison the typewriter keys with "some juice of the cursèd hebona" and, barring his ability to trick Rosenbaum into testing out the typewriter, he will challenge him to a duel, "casually graze his rapier over the poisoned typewriter keys and deliver a palpable hit!" His tertiary plan lurches from poisoned cup in *Hamlet* off into farce, and Danny Kaye territory: "We'll put the pellet with the poison in the vessel with the pestle!"[28]

As the play draws to a close, Swift continues to plot. Milton, the apologist for Rosenbaum (Swift: "How come you're always so goddamn ready to justify the ways of Rosenbaum to the apes?"[29]) even as his namesake justified the ways of God to men, continues to work, because "the man is going to want his pages."[30] And Kafka quietly begins to type *Hamlet*, beginning from Act One, Scene One.[31] Perhaps Kafka's *Hamlet*, like Pierre Menard's *Quixote*, will actually be superior to the original; the play ends at this point, so the audience is left with its speculations. Along the way, though, *Words, Words, Words* has laid bare many of the assumptions that the other typing monkeys cannot see, or choose to ignore, or refuse to articulate. See no evil, hear no evil, speak no evil.

Chapter 23

Barnyard Politics

Typing monkeys are one thing. Typing cows are quite another. No one ever set out to prove anything by using typing cows as an example. Until now, that is.

"Farmer Brown has a problem. His cows like to type." So begins Doreen Cronin's bestselling children's book, *Click, Clack, Moo: Cows That Type.*[1] A decidedly annoyed-looking Farmer Brown is exposed for several pages to the aural evidence of the cows' activities: "Click, clack, moo. Click, clack, moo. Clickety, clack, moo." "Moo" (or "Mu") has always been an aural signifier of emptiness; it is the name of a lost continent, and, according to the often inaccurate but always entertaining *Principia Discordia* (the psychedelic bible of the Discordian religion[2]), not only the sound that cows make, but also the name of the Chinese ideogram for "No-thing" (and, in the *Principia*'s mock koan, "Upon hearing this, absolutely nobody was enlightened").[3] Later in Cronin's book, when the cows hold an emergency meeting, "All the animals gathered around the barn to snoop, but none of them could understand Moo."[4] "Moo," then, is always and only the

Click, clack, moo.
Click, clack, moo.
Clickety, clack, moo.

The cows composing their list of demands.

sound that cows make, but the noise of typewriting always indicates active thought, if not always on the part of the typist.

But this is a children's book, and while the sonic component is important, the visuals are primary and convey the most information. The material quality of the type provides information to the reader that is inaccessible to Farmer Brown; the fact that the "clicks" and "clicketys" and "clacks" are rendered in a thick version of a Courier-style typewriter font is incontrovertible evidence to us that the cows are typing pages before the farmer accepts it. "At first, he couldn't believe his ears. Cows that type? Impossible!"[5] The sight of typed text is also what finally convinces the farmer, who finds a note nailed to the barn door (which begs the question of who did the hammering – but presumably, if cows can type, they can also hammer a nail):

Dear Farmer Brown,
The barn is very cold
at night.
We'd like some electric
blankets.
Sincerely,
The Cows[6]

It's no William Carlos Williams poem, but its brevity connotes the same kind of elegant immediacy. Unlike the monkeys, the cows are generative typists. The monkeys are enthralled to the system of dictation, and, even in the case of the Ives play, when it's clear that the monkeys have creative urges and desires other than those of the voices and powers that compel them to repeat the texts of others, their own compositions are regarded as nonsense. The cows have real desires born of real material needs, and the typewriter provides the means for them to transcend their dumbness and articulate those needs. Once again, typewriting produces truth (the cows sign their note "Sincerely," because they make this request in good faith).

Farmer John is not pleased with this turn of events: "It was bad enough the cows had found the old typewriter in the barn, now they wanted electric blankets!"[7] When he refuses to grant their request, the cows go on strike.

Every time the cows begin typing, more demands are forthcoming. Soon the hens and the cows form a coalition, with the hens also ceasing production until they too receive electric blankets.[8] Farmer Brown's barnyard has taken on the dimensions of a classic organized labour-versus-management dispute.

Emboldened by their initial foray, the cows and hens refuse to capitulate, and their communiqués become more terse.

Until this point, Farmer Brown has done a lot of shouting and gesticulating, but there has been no real communication between the disputing parties. In order for that to happen, he too must become a typist.

Duck, supposedly a neutral party to the dispute, bears the farmer's typed note to the cows.[9] After an all-night meeting, the duck returns to the farmer with another typed note. If the farmer leaves a stack of electric blankets outside the barn door, they will send the farmer their typewriter via the duck.[10] This is an interesting turn of events for a number of reasons. At no point did the farmer explicitly demand that the cows cede their typewriter. Unlike the electric blankets, the abandoned manual typewriter creates no significant drain on his resources or production costs. And yet the cows are willing to give up the only device that gives them a voice in exchange for material comfort (with no regard for the possible health hazards posed by electric blankets, either). The farmer decides "this was a good deal." And why not? If the cows yield the typewriter, it will be impossible for them to make further demands or foment unrest in the barnyard. He delivers the blankets, and the reader sees the cows and chickens blissfully asleep beneath them, but mute for the rest of the book.

What the farmer does not realize, though it is obvious to us from the outset, is that the cows are not the problem. It is *typewriting* that creates the possibility of the articulation of hitherto suppressed dissatisfaction and desire. It is *typewriting* that is the threat to the established order, suggesting, once again, that there is still a dictating voice at play in the circuit, even if it appears to be completely suppressed, sending out the same messages regardless of amanuensis. The farmer waits all night for the duck to return the typewriter, but in the morning he receives something else instead: a note from the ducks requesting a diving board for the pond.[11]

The last text in the book is once again the onomatopoeic sound of the typewriter in action, but "Click, Clack, Moo" becomes "Click, Clack, Quack" (adding the benefit of rhyme to the composition, not unlike the typing of Ives's monkeys). Almost virally, typewriting becomes the vehicle of liberation.

The ending of the book remains ambiguous, however. The last page bears no text, only the image of a duck's tail and legs above a splash in the pond, with a diving board in the background. Presumably, the farmer has acceded to the ducks' wishes, but the text leaves no information as to the whereabouts of the typewriter. Did the ducks, too, have to yield their ability to communicate? Did another species benefit from receiving the typewriter, as they did? So many questions . . .

Duck, it turns out, has political aspirations beyond his role as union arbitrator. Like Swift the monkey, he is uncomfortable with his assigned chores and yearns to overthrow his keeper. In *Duck for President*,[12] the sequel to *Click, Clack, Moo*, he does exactly that, holding, running, and winning an election to replace Farmer Brown. When he discovers that running the farm is more work than he wants to do, Duck becomes an embodiment of the Peter Principle, first running for and then becoming Governor, then President. At the end of the book, after he has given up on all of these jobs as too stressful, he returns to the farm to write his memoirs – on a word processor. Here, finally, is the answer to the location of the old typewriter from the barn; it's visible in the garbage can beside Duck's desk.

Though Duck has moved on into the final chapter of this book – typing after the typewriter – there are other animals that have not.

Chapter 24

Not an Especially Bright Dog

Elisabeth Mann Borgese, one of Nobel prizewinner Thomas Mann's six children, was a multilingual citizen of four countries, a science fiction writer, a pianist, a drafter of international oceanic law, and an environmental activist who lived in Atlantic Canada (chiefly Halifax) from 1978 until her death in 2002.

Borgese's own writing career parallels the entire cycle of the history of typewriting. As a young woman, she studied piano with Vladimir Horowitz at a Zurich conservatory. After marrying the Italian author, political scientist, and literary critic Giuseppe Antonio Borgese in 1938, she "let the piano take a back seat" to her marriage and taught herself shorthand in both English and Italian, along with other secretarial skills, "to be a more efficient secretary to her husband," who died in 1952.[1] After her father died in 1955, Borgese began writing her own stories, plays, and operas, as well as making significant contributions to international marine law.

Among her other accomplishments, Borgese taught her English setter, Arlecchino, or Arli, to type, a process she details in *The Language Barrier: Beasts and Men*.[2] Arli had a vocabulary of sixty words and seventeen letters; Borgese conceded, "He isn't an especially bright dog"[3] (as if to confirm this contention, German researchers in 2004 were working with a nine-year-old border collie named Rico who had a vocabulary of two hundred words[4]). "[Arli] could write under dictation short words, three-letter words, four-letter words, two-letter words: 'good dog; go; bad.' And he would type it out. There were more letters but I never got him to use more than 17," Borgese said.[5]

In October 1962, Borgese began training her four dogs in "school" for fifteen minutes a day. Beginning with simple conditioning involving edible rewards placed inside cups covered with plain white saucers, Borgese first taught the dogs how to make binary choices. Then, using the visual symbol vocabulary Professor Bernard Rensch had developed for use with elephants, she taught the dogs to distinguish between the eighteen different designs.[6]

As three of the dogs began to reach their limits for learning, Borgese focused exclusively on Arli. By January 1963, Arli could count to four, and could distinguish between the words CAT and DOG. In another month, he could count to six, and recognized six words. Borgese then began to break down the words into their constitutive letters and taught Arli to recognize the proper sequence of the letters, even when they were scrambled.[7] At the end of March 1963, Borgese presented Arli with what she called his first "keyboard": the letter sequence LATIRC. After hundreds of attempts, Arli eventually learned to produce the words ARLI, CAT, and CAR from this sequence. It was only then that Borgese confronted him with an actual typewriter.[8]

The Olivetti corporation donated a Lexicon 80 electric typewriter to the experiment, which Borgese had modified. An electrician attached a superstructure to the machine that connected the machine's original keys to a set of larger ones – two-and-a-half-inch white discs that looked like the saucers with which Arli was so familiar. Arli was not a QWERTY typist; he had his own unique keyboard:

Arli's retrofitted typewriter. The simplified keyboard layout resembles that of an Italian typewriter, but Arli typed in English.

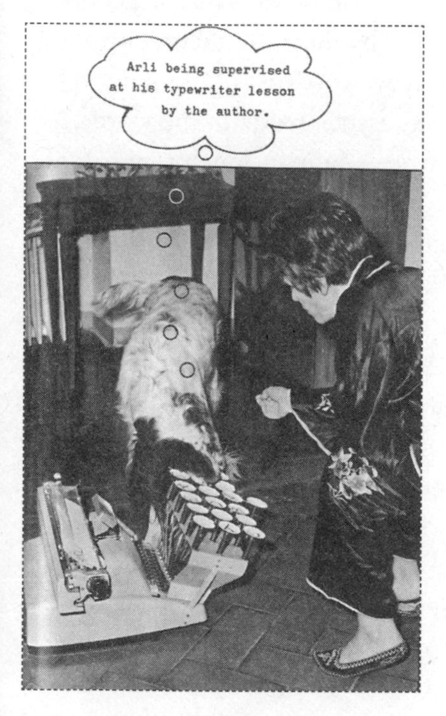

Elisabeth Borgese dictating to Arli.

A close-up of Arli typing.

Q Z E R T U I
A S D F G O P
C V B N H L M
(SPACE)

Borgese had initially placed a magnifying lens over the copy, but removed it when it became obvious that "there was no way of drawing [Arli's] attention to the written sheet or of connecting the finished product with the activity of typing."[9] Whatever else he managed, Arli would never be a touch typist.

In July of 1963, Arli used the typewriter for the first time. All the keys were blank, except for the A key. With a piece of raw hamburger as incentive, Arli first learned to ignore the humming of the machine, then to nose the key to type an A on command. Borgese pronounced each letter as Arli typed it, which meant that she could eventually introduce new words to Arli by dictating them. Borgese notes that this process of acquiring new words was "mechanical" on Arli's part; "No meaning at all was associated with the words." In this respect, Arli is no different from a human touch typist, for whom typing is also an automatic process involving internalized sequences of letters.[10] Within a couple of weeks, Arli had learned to write all the words contained in his LATIRC "keyboard." Within six months, Arli could type twenty words.[11]

Like any other neophyte typist, Arli made a lot of typos. Perhaps a third of these errors were classifiable, and broke down as follows. About per cent of his errors were simple keying errors, where he would nose a neighbouring key instead of the correct one. Another 3 per cent of his errors resulted from forgetting which sequence he was typing and beginning to type another with a similar middle component (for example, Borgese speculated that because the letters AR appear in both CAR and ARLI, Arli might sometimes type CARL as a result). The third category of errors (about 22 per cent) is interesting, because they resulted from instances where Arli *did* associate meaning with words that excited him. "When asked, 'Arli, where do you want to go?' he will unfailingly write CAR, except that his excitement is such that the 'dance' around the word becomes a real 'stammering' on the typewriter. ACCACCAAARR he will write. GGOGO CAARR."[12]

After several years of practice, Arli could type entire dictated sentences without error. Borgese began to wonder: could Arli become a generative typist? One incident that led her down this line of inquiry is the best-known Arli anecdote. Arli had been suffering from intestinal problems following a long flight, and was indifferent to his typing lessons. Borgese began to dictate: "G-o-o-d d-o-g g-e-t b-o-n-e," but Arli rolled around on the floor and generally behaved in a doglike manner. Finally, he stretched, yawned, and typed "a bad a bad doog." When writing about this incident, Borgese invokes our old friends the typing monkeys, but only to dismiss the event as pure randomness: "Of course, it may have been pure chance: not of the kind that will enable a million monkeys to type, in a million years, the Bible," which would have required a probability of 1:17. What she argues instead is that because the phrase "a bad dog" was one of the dozen or so sequences of characters Arli had "memorized," just like a touch typist, that the event was an occurrence with a likelihood of about one in twelve. Borgese remarks that some of his other common phrases, such as "Arli go bed," would have appeared just as meaningful under the circumstances.[13]

In March 1965, Borgese began experimenting with letting Arli type without dictation. What is interesting is that by this point the texts that Arli produced were immediately recognizable as "poems," not only to Borgese but to an unnamed "well-known critic of modern poetry" to whom she showed the texts. The critic wrote, "I think he has a definite affinity with the 'concretist' groups in Brazil, Scotland, and Germany [and an unnamed young American poet] who is also writing poetry of this type at present."[14]

For Arli, though, like all other typing poets, generative typing took a heavy toll. If left too long in front of a typewriter without instruction, Arli began to whimper and whine and hit the machine with his paw. Borgese closes her chapter on Arli with the following bit of ventriloquism: "'How should *I* know what to do?' he seems to say. 'Dictate!' For heaven's sake *dictate*!"[15] This is a complex utterance: the dictator mimicking the voice of the amanuensis, commanding from a position of subordination that dictation begin. If Arli truly *did* have volition, the relationship might be described as masochistic, where the amanuensis/bottom is contracting for a disciplinary scene of their own construction, but the projection of this fantasy onto the dog only makes him seem more abject.

There *is* one more typing animal that I'd like to discuss, one that ties this narrative back to both Kafka and Cronenberg's *Naked Lunch*. In keeping with the development of this chapter, there is also an element of abjection, a little dictation from beyond the grave, and some lessons about both the limitations and the benefits of the typewriting grid. May I present Archy, the typing cockroach.

Chapter 25

under difficulties semi colon

In 1912, a young journalist for the New York *Evening Sun* named Don Marquis began writing his own daily column, "The Sun Dial." Producing a daily column is arduous work that requires patience and discipline. While a creative and prolific individual, Marquis was neither patient nor disciplined, and as a result often found himself pressed for material.[1] In 1916, he hit upon a brilliant solution.

That solution begins with Marquis relating the following anecdote to his readers:

> Dobbs Ferry possesses a rat which slips out of his lair at night and runs a typewriting machine in a garage. Unfortunately, he has always been interrupted by the watchman before he could produce a complete story.
>
> It was at first thought that the power which made the typewriter run was a ghost, instead of a rat. It seems likely to us that it was both a ghost and a rat. Mme. Blavatsky's ego went into a white horse after she passed over, and someone's

> personality has undoubtedly gone into this rat. It is an era of belief in communications from the spirit land.[2]

There you have it: typists are abject vermin. But more than that, perhaps even because of that, in combination with a typewriter, they provide a channel for a dictating voice from elsewhere, some "power which made the typewriter run," for miraculous as the fact of typing vermin might be, it evidently cannot account for the production of writing on its own.

Moreover, the phenomenon is not limited to Dobbs Ferry's rat. Marquis comes into his office early one morning only to find, to his considerable surprise, "a giant cockroach jumping about on the keys":

> He did not see us, and we watched him. He would climb painfully upon the framework of the machine and cast himself with all his force upon a key, head downward, and his weight and the impact of the blow were just sufficient to operate the machine, one slow letter after another. He could not work the capital letters, and he had a great deal of difficulty operating the mechanism that shifts the paper so that a fresh line may be started. We never saw a cockroach work so hard or perspire so freely in all our lives before. After about an hour of this frightfully difficult literary labor he fell to the floor exhausted, and we saw him creep feebly into a nest of the poems which are always there in profusion.[3]

This is Archy, or archy, depending on which side of the fierce debate over the orthography of his name you stand. As it requires the force of his entire body simply to make an impression on the key, he is usually unable to type any of the keys that require a shift, including upper-case letters. Thus, he signs his own name "archy." As E. B. White observes in his introduction to a major Marquis

One of George Herriman's many cartoons of Archy at work.

collection, "Archy . . . was no e. e. cummings," which connotes not only the inferior quality of his verse, but that Archy typed his name the way he did out of necessity rather than affectation. White then adds that Don Marquis himself was in the habit of capitalizing Archy's name when referring to him.[4] In one famous piece, "archy protests," the cockroach implies that his all–lower case style is a signifier of his implicit protest against his appalling working conditions *and* the philistinism of critics who fail to note those conditions:

say comma boss comma capital
i apostrophe m getting tired of
being joshed about my
punctuation period capital t followed by
he idea seems to be

that capital i apostrophe m
ignorant where punctuation
is concerned period capital n followed by
o such thing semi
colon the fact is that
the mechanical exigencies of
the case prevent my use of
all the characters on the
typewriter keyboard period
capital i apostrophe m
doing the best capital
i can under difficulties semi colon
and capital i apostrophe m
grieved at the unkindness
of the criticism period please
consider that my name
is signed in small
caps period
archy period[5]

Archy normally does without punctuation entirely, but his decision here to spell out the punctuation is ingenious. Not only does it give him access to the characters he cannot normally employ, but he can also emphasize the arduousness of the act of typewriting itself.

In any event, Archy is a "vers libre bard" (yet another typing poet) who "died and my soul went into the body of a cockroach."[6] As if the struggle of typing poetry itself were not enough, a struggle that Archy repeats night after night "for you / on your typewriter" (he too is in servitude from the outset, referring to Marquis always as "Boss"), a rat named Freddy (another reincarnated poet) habitually first scorns Archy's poetry, then eats it, erasing an entire night's work.[7]

Archy is, of course, a thinly veiled analogy for the people that White refers to as the "thousands of poets and creators and

"He would climb painfully upon the framework of the machine and cast himself with all his force upon a key, head downward . . ."

newspaper slaves"[8] that fill all of the column inches of the daily papers of the modern world with typewriting, especially Marquis himself. When he introduced Archy, White writes, "Marquis was writing his own obituary notice."[9] "[Marquis] was never a robust man," observes White, adding that he "usually had a puffy, overweight look and a gray complexion."[10] Marquis's writing career and home life were anything but smooth. After nearly exhausting himself on his column, he switched to playwriting, and, after making a small fortune on a play based on another of his characters, The Old Soak, lost it all on his next play, about the Crucifixion (unlike Mel Gibson, Marquis was a sceptic). A dalliance with writing for Hollywood left him bitter and vituperative. Most difficult of all, Marquis suffered the deaths of two children and two wives in less than fifteen years, and died penniless, sick, and blind after a series of strokes in 1937.[11]

Archy was, as White observes, "the child of compulsion, the stern compulsion of journalism,"[12] meaning that not only did he reflect the material quality of Marquis' writing life, he also solved a number of problems for Marquis on a technical level. Because Archy wrote in free verse, Marquis suddenly had licence to write very short lines, lines that did not have to fill the whole width of his oppressively wide column. Runover lines were no longer an issue, because every line was broken. Without upper case or punctuation to worry about, there was less to copy edit. In addition, Marquis could rely on the logic of rhyme and other paratactic structures to power his writing forward when normative syntax failed him; doggerel is always easier to produce than incisive journalistic analysis. "Thanks to Archy," White writes, "Marquis was able to write rapidly and almost (but not quite) carelessly."[13]

That relative freedom allowed Marquis to produce, via Archy and his friend Mehitabel the cat (a reincarnation of Cleopatra), an enduring body of literature. Since its first appearance in 1927, *Archy and Mehitabel* has been in print continuously, perhaps because it delineates the agonism of typewriting in a fashion that is lighthearted and poignant by turns. Archy does have his small moments of triumph:

> I THOUGHT THAT SOME HISTORIC DAY
> SHIFT KEYS WOULD LOCK IN SUCH A WAY
> THAT MY POETIC FEET WOULD FALL
> UPON EACH CLICKING CAPITAL
> AND NOW FROM KEY TO KEY I CLIMB
> TO WRITE MY GRATITUDE IN RHYME[14]

. . . but even on this singular occasion, which Archy marks by putting in the extra effort to rhyme his composition, a sudden capricious attack by Mehitabel unlocks the shift and knocks archy "right / out of parnassus back into / the vers libre slums i lay / in behind the wires for an hour after."[15] Mostly, Archy's lower case world is a kind of prison, and even his dreams of machine-aided transcendence take the form of a plea:

> . . . say boss please lock the shift
> key tight some night
> I would like to tell the story of
> my life in all capital
> letters
>
> archy[16]

It never happens again.

Chapter 26

Bourgeois Paper-Bangers

Meanwhile, back in the world of human typists, people are becoming increasingly annoyed with their typewriters.

The narrator of Tom Robbins's *Still Life with Woodpecker* faces problems similar to Cronenberg's William Lee: his typewriter has a mind of its own. *Still Life with Woodpecker* stages itself as the narrator's struggle to wrest "the novel of [his] dreams"[1] from his newly purchased state-of-the-art Remington SL-3, a typewriter that exudes an aura of dictatorial power: "[It affects] even by candlelight, the suspecting, censorious glare of the customs inspector or efficiency engineer. It appears to be looking over my shoulder even as I am looking over its."[2] The set of cultural rules that determine the logic of the office typewriter are entirely unsuited to the task that the author sets it; he scornfully refers to it as a "bourgeois paper-banger."[3]

He fantasizes about a number of machines more suited to his project, "say, a Remington built of balsa wood, its parts glued together like a boyhood model; delicate, graceful, submissive, as ready to soar as an ace," or "better, a carved typewriter, hewn from a single block of sacred cypress; decorated with mineral pigments, berry juice, and mud; its keys living mushrooms, its ribbon the long iridescent tongue of a lizard," and "a typewriter constructed of tiny seashells by a retired merchant sailor, built inside a bottle so that it can be worked only by the little finger of the left hand of a right-handed person."[4] The narrator notes that "I'm not so far gone that I expect technologists to be interested in designing machines for artists"[5] – such fantastic machines represent a reversal of the dominant logic of typewriting, one that would tip the relationship in favour of the generative typist.

Therefore, the narrator makes an attempt to swerve away from the oppressive logic of his machine by physically altering it, painting it a red "as ruddy and indiscreet as a plastic sack full of hickeys"[6] in order to enable it to "cope with letters, words, sentence structures with which no existing typewriter has had experience."[7] He goes on, only to discover that "the Remington warranty doesn't cover 'typing of this nature,' whatever that might mean."[8]

Robbins's narrator is ultimately unsuccessful in his attempt to depart from the typewriter's logic while continuing to utilize it. He finally abandons the machine entirely in favour of a return to what the typewriter's boosters have always called "pen slavery" in the final pages.[9] One phrase in the text points toward our next major issue. "If novelists got wooden typewriters, poets would demand that theirs be ice," he types.[10] Ice connotes several things, among them, ephemerality and slipperiness. More than anything else, all of the discipline in typewriting strains toward its own transcendence, a transcendence through speed.

Part

Acceleration: Typewriting and Speed

"THE MACHINE HAS SEVERAL VIRTUES I BELIEVE IT WILL PRINT FASTER THAN I CAN WRITE."

– Mark Twain, in a letter to his brother

Chapter 27

Rail Road Test

In April of 1929, Northrop Frye took the train from Moncton to Toronto for the first time. At his side was a wooden case containing a borrowed Underwood typewriter. He was coming not to attend school, but to participate in a typing contest.

At fifteen years of age, as an award for highest standing in English at Aberdeen High, Frye received a scholarship for three months of stenographic training at Success Business College. In his biography of Frye, John Ayre notes that "presumably because of his facility with the piano" Frye was a nimble typist, working at speeds approaching 70 words per minute. As a promotional gimmick, the owner of the business college offered Frye a free trip to Toronto to participate in Underwood's national typing contest. Ayre also notes that "this was the heyday of the typing contests, which, like all Jazz Age extravaganzas, were conducted with high publicity and consummate bad taste."[1]

Bathed in the bright lights of no less a venue than Massey Hall, Frye, registered in the novice class, came in second – barely. He typed an average of 63 corrected words per minute, edging a lead over a young woman who managed 62 words per minute, but falling

The young Northrop Frye: number two always tries harder.

significantly short of the first-place winner, who had managed 69 words per minute on her machine.[2] After returning to Moncton for the summer, Frye did some office work, then spent more time at Success honing his typing skills. Soon, Frye was typing at top speeds of 85 words per minute.[3] An international typing contest brought Frye back to Toronto the following year. Ayre doesn't mention how he did this time around, so Frye's other achievements will have to suffice.

Half Measures

As I noted earlier regarding time and motion studies, an increasingly fine control over the partitioning of time is part and parcel of systems of discipline. What I haven't discussed in detail is that one of the *goals* of those studies is an increase in speed. Once time can be measured, it becomes something that can be saved; the more time becomes measurable, the more of it can be recovered, in order to be spent in a way that ensures maximum productivity and therefore profitability.

Even the earliest accounts of writing machines express an interest in rapid as well as clear writing. For written dictation to supplant spoken dictation, the speed of a writing machine had to be at least as rapid as that of speech itself. Friedrich Kittler pinpoints 1810 as the moment when the technology of impressing words onto paper finally begins to match the desired speed. In that year, the introduction of rotary presses and continuous forms created a system where "types fall, through the touch of the keys, into place almost as quickly as one speaks."[4]

Almost. But any system that merely came close to the speed of speech would simply not do. As *The Story of the Typewriter* suggests with its typical boosterism, shorthand and stenography needed to be abandoned in favour of typewriting for precisely this reason:

> It is obvious that shorthand, even as perfected by phonography, would have been restricted, without the typewriter, to a limited field of usefulness. As a time saver, shorthand is clearly a half measure, and, so long as the art of transcribing notes in long hand could be done only at pen-writing speed [which the book identifies elsewhere as approximately 30 words per minute[5]], the swiftest shorthand writer could render only a partial time-saving service. In the days before typewriting, it would have required more than one stenographic secretary to free the busy executive from the bondage of the pen.[6]

In 1840, the speed of handwriting began to seem almost tectonic when compared to another invention patented in that year: the telegraph.[7]

For typewriting, time was short from the outset. Approximately fifty years before the typewriter was perfected for commercial use, Joseph Henry demonstrated, in 1830, that long-distance, networked communication was possible, when he used electromagnets to cause a bell to ring at the far end of a mile of wire. Five years later, Samuel Morse was expanding on Henry's work. He had already made the jump from aural signalling to written signalling by attaching a marker to the end of his electromagnet. Long and short electrical pulses thus became long and short dashes on a sheet of paper: the first Morse Code. Morse was holding public demonstrations in 1838, and secured his patent two years later. Compared to the 30 words per minute of handwriting, a trained telegraph operator could send messages at 50 words per minute.[8]

Typewriting clearly had some catching up to do. The only problem was that no one knew how to type yet.

Mark Twain on the Burning Deck

Mark Twain was supposedly the author of the first typewritten manuscript submitted to a publisher. Twain himself boasted of it frequently, though he got the details wrong just as frequently. In 1905 he wrote, "I will now claim – until dispossessed – that I was the first person in the world to *apply the type-machine to literature*. That book must have been *The Adventures of Tom Sawyer*."[9] (Actually, it was *Life on the Mississippi*[10]; Twain's error here has rippled down through the scholars up to and including Kittler.[11]) The story of how Twain came to have a typewriter in his possession is less well known, but is worth relating, because it bears on the question of dictation as well as on how typing speed became a marketing tool, even before touch typing had been invented.

Twain begins his anecdote by reflecting that while he had dictated to a typewriter before (indicating that he was also dictating this essay), he had never dictated autobiography. Nevertheless, he proceeds, "it goes very well, and is going to save time and 'language.' "

What Twain has in mind by "language" is evidently profanity, as he adds "the kind of language that soothes vexation." Both vexation and the cursing that ameliorates it are, in this formulation, excessive and unnecessary. For Twain, typewriting is an act of conservation of precious resources that could be better deployed elsewhere.

Twain's use of the word "typewriter" – twice in his first two paragraphs – is also interesting. What he means, however, is not easy to determine, because he does not use the word again in the essay. When he refers to the mechanical device itself, he uses "type-machine"[12] or "machine"[13]; when he refers to a typist or amanuensis, he uses "type-girl"[14] (he also uses the term "machinist" to refer to a typist[15]). All of these other terms appear several times. While it's possible that Twain is using "typewriter" as a synonym for one or the other, it's unclear which it might be. It's equally likely, since he is speaking of the process of writing-as-dictating-to-the-typewriter, that he is referring to the entire assemblage.

Between the time that Twain is dictating this passage and the time he first saw the typewriter, in 1873, the year the Remington Type-Writer No. 1 was issued, the number of typists in the United States increased from around 150 to more than 110,000.[16] Twain himself observes that, in 1873, both the "type-machine" and the person owning it were curiosities, but that the situation had since reversed; the aberration had become the mundane.[17]

Twain's first encounter with the "type-machine" is marked by the same kind of vexation that his later dictation to the "typewriter" soothes away. While lecturing in Boston, Twain and his friend Petroleum V. Nasby (the pseudonym of journalist and satirist David Ross Locke[18]) see the "machine" in a shop window. Twain, an inveterate lover of gadgets, is entranced, and they go in to request a demonstration. The salesman claims it can produce 57 words per minute, and when Twain and Nasby express incredulity, "he put his type-girl to work" (she appears at this point with no prior warning, as if she were a component of the machine that had been implicit yet invisible all this time). Twain and Nasby time the demonstration and are "partly convinced" but require repetition for verification. "We timed the girl over and over again," Twain writes; their incredulity translates into an indefinitely long series of repetitions of the event. After each round, Twain and Nasby pocket the narrow slips of paper that the demonstration produces "as curiosities."[19] At this point, the conjunction of inorganic machine and organic type-girl is still an aberrance, and so are the documents that it produces.

Excited, Twain immediately pays the full asking price of $125 and returns to his hotel. When he examines the typed slips more closely, Twain is "disappointed" to realize that all the samples are identical: "The girl had economized time and labour by memorizing a formula." There is nothing here that is spontaneous, free, or irreplaceable. When Twain returns home and his own machine finally arrives, he has already relegated it to the status of "toy"; it is little different to him than the clockwork boys and other writing automata of the previous century. Moreover, the toy is monstrous: Twain writes later that "that early machine was full of caprices, full of defects – devilish ones." It exerts a corrupting influence not only on the morals of Twain himself, but on others he attempts to foist the machine upon.[20] All he thinks to do with it is to mechanically attempt to duplicate the type-girl's feat (and he does only indifferently at that). He types and retypes the first line of Felicia Dorothea Hemans's poem "Casabianca" – "The boy stood on the burning deck" – to the point where he "could turn out that boy's adventure at the rate of twelve words a minute; then I resumed the pen for business, and only worked the machine to astonish inquisitive visitors. They carried off reams of the boy and his burning deck."[21]

Eventually, Twain "hired a young woman, and did my first dictating." The substance of that dictation was "letters, merely," a kind of writing he considered inferior, and the "machine," being a Model 1, only rendered them in capitals "and sufficiently ugly."

The only use Twain mentions explicitly for such a letter in this essay is a deliberate foiling of the request by Mr. Edward Bok (editor of the *Ladies' Home Journal* and himself a former amanuensis) for "a whole autograph *letter*"[22] by responding "signature and all"[23] in full typed gothic capitals. For Twain, this "machine" only produced a kind of writing that was *not* writing; he comically reproaches Bok that "it was not fair to ask a man to give away samples of his trade."[24]

It is at this point that Twain reaches "an important matter" – the aforementioned claim that "I was the first person in the world to *apply the type-machine to literature.*" Yet he prefaces this claim with "In the year '74 the young woman copied a considerable part of a book of mine *on the machine.*" As per Leah Price's argument, Twain's typist has done the actual typing, but abrogated her right to have her signature on the first piece of typed literature.

Actually, this erasure of a "first" by a woman typing happens not once but twice in Twain's history. The following string appears across the top of Twain's first typed letter:

> BJUYT KIOP M LKJHGFDSA:QWERYUTIOP:,-98V*6432QW RT
> HA[25]

These characters were written not by Twain, but by his daughter Susie. Arguably, Susie Clemens is the author of the first avant-garde poem composed on a typewriter, but, like the texts produced by Ives's typing monkeys, it is usually relegated to the status of marginalia and error. At the very least, it traces, in something like the fashion that Olson imagines, an inquisitive hand exploring the order of the keys on the keyboard. LKJHGFDSA:QWERYUTIOP is a sequence tracing the movement of a typing hand from the right side of the keyboard, where the first two clusters of letters occur, leftward across the entire middle row, then back toward the right across the top row, in the famous sequence that gives the keyboard its name. Susie's name, unlike that of the machine, is not here; all we have by way of her assertion is the triumphant "HA" at the end.

After he tires of the burning deck, Twain attempts to give the machine away again and again, but like any other cursed object, it always returns back to him. Finally, Twain gives the typewriter to his coachman, who trades it to "a heretic for a side-saddle he could not use," and Twain finally loses track of the machine.

The point where the true event occurs, the rethinking of both the machine and the performative event itself, is at the beginning of Twain's essay. The piece he has just composed is free, spontaneous, irreplaceable, and so on; it demonstrates, by the machine, that a transformation has occurred. Twain is now typewriting, without subordinating either the event or the machine.

And yet. There is still a Type-Writer Girl or amanuensis in the circuit, another woman who types silently, spectrally. It is telling that when Twain's coachman, the penultimate owner of the typewriter in the essay (though Twain now refers to it as "the animal"[26] rather than "the machine"; in typewriting, the organic and the inorganic are now inextricably mixed), tries to rid himself of the typewriter, the only barter object capable of making the infernal machine disappear is a sidesaddle someone offers him in trade for it, a pseudo-useful device designed to enable women to use a particular kind of technology, but not to its full capabilities. So there she sits, Twain's anonymous amanuensis, typing "Mark Twain" in place of her own name. "Mark Twain," of course, is already a pseudonym, not a writer but an author-function, indicating the presence of at least two people.

There is little comfort for either of Twain's type-girls that he and Nasby rationalized their first disappointing experience with the notion that, like the first billiard player, the first type-girl could not be "expected to get out of the game any more than a third or a half of what was in it." In the next sentence, Twain writes, with the benefit of hindsight, "If the machine survived – *if* it survived – experts would come to the front, by and by, who would double this girl's output without a doubt. They would do one hundred words a minute – my talking speed on the platform."[27]

Chapter 28

Ten Fingers Talk

When new technologies and new ways of thinking first appear, the uses their inventors intend for them usually have little to do with what people ultimately make of them.[1] Put another way, as Bruce Bliven observes in *The Wonderful Writing Machine*, "By today's standards, no one knew how to type for fifteen years after the typewriter came onto the market."[2] The first appearance of the word "touch" as a modifier for "typing" is in the 1889 publication *A Manual of Practical Typing*, by Bates Torrey.[3] Like most neologisms, it enters the language long after the phenomenon it names.

Like the typewriter itself, touch typing has only relative beginnings. The idea seems to have occurred, more or less simultaneously, to two people who didn't know each other and had probably never even heard of each other: Mrs. L. V. Longley, proprietor of Longley's Shorthand and Typewriter Institute, and Frank E. McGurrin, official stenographer for Federal Court in Salt Lake City, Utah.

Mrs. Longley is indisputably the author of the first document arguing that typists should employ all their fingers in the process of typing rather than hunt and peck with two or four fingers. By 1881, she was referring to her process as the "All Finger Method," and in 1882 published a book outlining a curriculum that employed it, called *Remington Typewriter Lessons*.[4] The All-Finger Method was not touch typing proper, since it did not concern whether the typist was to gaze on the copy, the keyboard, or the source text.[5] Nevertheless, the text's own copy acknowledges that it is the originator of a new discursive practice, "a system of fingering entirely different from that of other authors and teachers."[6]

Bliven reports that Longley "had been teaching her students the method for some time before she got it down on paper."[7] The confusion about how long "some time" actually amounts to is, ironically, due to someone's typo. Authorities agree that the professional reception of Longley's system was far from positive, citing a negative editorial from the *Cosmopolitan Shorthander*, a trade magazine: "Unless the third finger of the hand has been previously trained to touch the keys of a piano, we believe that it is not worth while to attempt to use that finger in operating the typewriter. The best operators we know of use only the first two fingers of each hand, and it is questionable whether a higher speed can be attained by use of three."[8] Bliven gives the date of this editorial as 1887[9]; Wilfred Beeching, on the other hand, in *Century of the Typewriter*, cites it as 1877.[10] The 1887 date is more likely and the transcription error likely Beeching's, as his book postdates Bliven's by twenty years, and the Beeching text often closely paraphrases Bliven's (albeit without many of Bliven's excesses).

Though he did not publish anything on the subject, Frank McGurrin, the other major contender for first touch typist, was reportedly using a personal all-finger system by 1878, on a Remington No. 1.[11] McGurrin's system was closer to a modern touch-typing system, because he had gone to the trouble of memorizing the keyboard, and could type blindfolded as proof.[12]

The ability to avoid looking at the keyboard or no more than occasionally at the copy (at the end of each paragraph or even less frequently) is the first of the three characteristics of touch typing today. The other two, as defined by typewriting scholar Hisao Yamada, are that the process of typing has been internalized to the extent that the typist does not have to voluntarily direct his or her fingers ("The sequential motion of the fingers is mostly the result of the subconscious cortical reflex of the cerebrum in response to visual stimuli from the manuscript"), and that the typist's minimal unit of motion is sequences of letters, not individual letters. Yamada believes that accomplished typists have internalized somewhere between two and three thousand of these sequences. According to Yamada's own research, it requires about four hundred hours of practice time to develop a sufficient number of cortical reflex patterns for a person to become a "skilled" typist and an additional six hundred hours to become an "excellent" typist (Yamada's terms are somewhat fuzzy; his criteria for an excellent typist is someone who is "truly smooth and relaxed").[13]

In other words, the fastest typists type reflexively, as part of an assemblage that processes strings of letters based on the criteria established by the human/machine interface that have little to do with syntax, meaning, or alphabetical order. Contemporary typists are capable of achieving speeds of over 110 words per minute, *provided that they not expend any effort on attempting to comprehend the meaning of the copy*. William E. Cooper's Introduction to *Cognitive Aspects of Skilled Typewriting* explains that today's typing training actively produces incomprehension of the typed text as a desired trait: "In contemporary secretarial schools, training emphasizes the inhibition of reading for meaning while typing, on the assumption that such reading will hinder high-speed performance. Some support for this assumption derives from the introspections of champion speed typists, who report that they seldom recall the meaning from the source material incidentally."[14]

Generative typists – that is, those typists whose dictating voice is physically absent from the scene of production, such as novelists, essayists, and other creative writers – can often achieve "relatively high speeds" with the aid of touch typing, "yet this skill seems to be acquired with considerable difficulty and only after touch typing from copy has become facile."[15] The Powers that Be, however, don't pay that much attention to generative typists, weirdos and misfits that they are . . . unless one of them inadvertently stumbles on something "useful." On the other hand, the struggles of those competing to make the dominant systems of discipline more efficient also make for excellent spectacle.

1, 2, 3, 4, I Declare a Typing War

The advent of modern touch typing was a truly combative moment: a typing "duel" between Frank McGurrin and Louis Taub. Like Mrs. Longley, Taub was a typing teacher from Cincinnati; unlike her, he championed a four-finger hunt-and-peck method *and* an alternate keyboard arrangement. Taub used a Caligraph, a seventy-two-key typewriter (six rows of twelve keys, one key per character, including both upper and lower case letters). Thomas A. Russo notes that "it was a light machine with a light touch and compared favorably in many of the initial speed typing contests against the Remington." An early ad for the Caligraph reads, in part: "Greatest Speed On Record! 126 WORDS PER MINUTE, EXCLUDING ERRORS! T.W. OSBORNE, winner of the International Contest at Toronto, wrote on the CALIGRAPH Writing Machine 630 words in five minutes, thus gaining for the CALIGRAPH the Championship of the World."[16] There was thus more at stake than the merits of a touch system; it was also a test of the QWERTY keyboard configuration against a popular rival.

In the decade that he'd had to perfect his system, McGurrin's reputation as a typist had spread to the point where he was touring the western United States, performing before large crowds. Part of McGurrin's performance was a challenge to race anyone who would contest his claim to be the world's fastest typist, with a stake of five hundred dollars to back up his boast. Taub contested; McGurrin challenged; and on July 25, 1888, they raced.[17]

The race itself had two components: forty-five minutes of dictation and forty-five minutes of copying from an unfamiliar text. McGurrin won both components, hands (and eyes) down. Due to his ability to work without lifting his eyes from the copy or his fingers from the keyboard, he typed even faster during the second phase of the race.[18]

The McGurrin versus Taub encounter sparked a thirty-five-year international craze for typewriter racing contests such as the one that brought Northrop Frye to Toronto. The birth of this sport occurred in approximately the same period as professional baseball[19]; unaccountably, the latter has enjoyed a much longer lifespan. Of course, the typewriter manufacturers stood to gain the greatest rewards from this craze, and most had their own stable of champion racers, led by coaches constantly seeking ways to shave valuable seconds off their typists' times. Championship typewriting, which requires *net* speeds of 120 to 150 words per minute (that is, after penalization of additional added seconds for each error) lays the machine's regime of discipline bare as it attempts to maximize performance, as Bliven details:

> If a contestant wrote more than seventy-six or less than sixty-one letters in a line, that was counted as a mistake. If a letter was not entirely legible, that was a mistake. If a letter failed to strike exactly in the middle of its space, or if the margin was not perfectly even, or if the escapement jumped a space, those were all errors; not to mention garden-variety mistakes, of course, like hitting a wrong letter, transposing, missing a word, or misspelling.[20]

The intense pressures on contestants necessitated a micromanagement of the physical space around them and their equipment, shimming chairs and tables to the appropriate height and staggering stacks of paper to enable more rapid paper changes.[21] The typewriters themselves were customized and fine-tuned to the preferences of their operators. Coaches like Underwood's legendary Charles E. Smith developed highly specialized ergonomic movements such as the "speed paper insert," a technique for removing a full sheet of typing and inserting a blank piece of paper into the machine's carriage while it was returning from left to right[22] – a task like trying to thread a moving needle. The desire to maximize performance also led to the adoption of odd ergonomic costumes: "Racers of both sexes used special racing typists' visors, long-beaked affairs covered in green cloth with their edges turned down sharply on both sides for about three inches. They cut off any glare from the overhead lights and, at the same time, acted like blinders on a race horse."[23]

Bliven's theories about the end of the "Golden Era" of typewriter racing are interesting. For one thing, largely due to the efforts of Charles Smith, the Underwood team consistently won a discouragingly large number of races, so the element of competition gradually began to wane.[24] The real culprit, however, was the very system of discipline that made the races possible. After every race, there was a mountain of paper for the judges to sift through. Each page in that pile had to be assessed on a character-by-character basis, with plenty of room for dispute due to the inevitable subjectivity of the judges. Spectators simply wouldn't wait around for the results.[25]

There are still typing contests today, and contemporary speed typists are scorchingly fast. The world's fastest typist is currently Barbara Blackburn, of Salem, Oregon, who has achieved the speed of 212 words per minute, regularly produces 170 words per minute, and can sustain 150 words per minute for 50 minutes at a time, with an error frequency of two-tenths of 1 per cent. The key to her success, she claims, is . . . the Dvorak keyboard. Blackburn's narrative is like that of the skinny kid on the beach who discovers the Charles Atlas method of bodybuilding and returns, muscle-bound, to wreak havoc on the bully who kicked sand in his face. As a high school student, she received a grade of "I," for Inferior, in her typing class. In her freshman year of business college in 1938, she discovered the Dvorak keyboard, and was soon typing at 138 words per minute.

This is not entirely surprising, as one of Leibowitz and Margolis's complaints about the Dvorak Navy studies was that there was no way to tell whether the results would hold true for skilled or expert typists as well as substandard ones. In any event, in a milieu of networked computers, the very notion of what constitutes efficient office work has changed drastically, as I'll argue in the final part of this book. Computer processor speed and massive storage capacities dwarf the benefits and in many instances obviate the need for rapid human typing. Merton Hollister, a Florida math professor, observes, "With boilerplate documents in the modern law firm, even I can type 4,600 words per minute. All I have to do is bring up the document, change a couple of words and print it out."[26]

The fickle attention of a public that once flocked to typewriter races has also drifted, as Blackburn's shoddy treatment at the hands of David Letterman demonstrates. One of the hallmarks of Letterman's brand of comedy is to turn earnest demonstrations of unusual or obscure skills into objects of derision. Evidently, speed typing now qualifies for this treatment. When Blackburn returned home to see her segment air on Letterman's show, she was chagrined to hear his opening remark: "No doubt Ms. Blackburn is a very nice lady, but she has to be the biggest fraud and con artist in the world." Once the benchmark of professional clerical skill, speed touch typing has become a joke.

Around the same time that mass culture suddenly lost interest in the novelty of typewriter racing, an entirely different class of writers began to appear who were very interested indeed in the literal relationship between typewriting and speed. However, if typewriting were an Olympic event, these writers would be the ones losing medals due to their experiments with what we refer to politely as "controlled substances."

Chapter 29

On the (Royal) Road

"Here I am at last with a typewriter," wrote the nineteen-year-old Jack Kerouac on October 13, 1941. "America is sick as a dog, I tell you. That's why, with my new typewriter and a lot of yellow paper, I am grown dead serious about my letters, my work, my stuff, my writing here in an American city (Hartford). I will talk all about life in the 20th Century, and about America's awful sickness and about some individual sickness I see in men and about good things in life, namely, books, October skies, other varied weathers, women, Johnny Barleycorn, the Love of Man, a warm roof."[1] From the beginning of Kerouac's recognition of himself as a writer, he conceived of the typewriter not only as an instrument for the production of truth, but also as a means of channelling and disciplining the relentless flow of images and ideas through his head.

That year, Kerouac was signing his name "Jack Kerouac, F. P." The "F. P." stood for "Furious Poet," and his anger at the sickness of America fuelled a furious rate of production. Kerouac claims to have produced on his rented Underwood typewriter something like two hundred short stories in about eight weeks. Paul Marion, the editor of Kerouac's juvenilia, writes that Kerouac "often filled the sheet of paper top to bottom and margin to margin. There was barely enough space to contain his overflowing mind, so he wrapped up the thought and punched in the final *jk* on the last line – and you imagine him reaching for another piece of paper to start a new line." Without explicitly saying so, Marion invokes Charles Olson to describe the young Kerouac's typing, saying he "uses the page as a compositional field."[2] The young Kerouac, from his eponymous perch "atop an Underwood," closes his Introduction to his frenzied work with an understatement: "You will also find, as you go along, that I am also making a drive toward a whole mess of other things."[3]

"When working on a manuscript," writes critic Tim Hunt, "Kerouac tended to push straight ahead."[4] But early in 1951, Jack Kerouac was blocked. He had been thinking about an ambitious project called *On the Road* for at least three years, and had made a variety of assays and false starts, punctuated by drug-addled cross-country car trips with Neal Cassady. Hunt argues that Kerouac's dilemma was formal: he needed a means of producing a book that was at once "tightly structured" and "objective" in order to convey the scope and importance of the ideas he wished to address, but that would also accommodate "the fluidity of the road experience."[5] In addition, after trying on and rejecting a number of philosophical approaches to the work, Kerouac had resolved "to write spontaneously and solely from inspiration rather than from preconceived notions about one's material."[6] Most of all, Kerouac sought to use typewriting in a manner that should be familiar by now: as an instrument for producing truth, in order to wrest poetry and prose "from the false hands of the false."[7]

Kerouac found his solution in a modified form of typewriting. The processes of discipline embedded in the machine provided the necessary tight structure. For any typist, the decision to write "from inspiration" concerns their ability to tune in closely to a particular dictating voice or voices. The rest comes from speed, in a variety of literal and figurative forms.

First, as Marcus Boon explains in *The Road of Excess*, Kerouac had access to a cheap and plentiful source of Benzedrine from over-the-counter nasal inhalers; he would remove the drug-soaked paper strips and ingest them with additional stimulants in the form of a coffee or cola chaser.[8] Second, Hunt claims that Kerouac could type 100 words per minute (though he doesn't state whether Kerouac achieved this velocity before or after the Benzedrine kicked in).[9] Kerouac's biographer, Gerald Nicoisia, mentions Kerouac's typing skills twice; even as a high school student, Kerouac was "already a speed typist [and] would sit at the machine at his father's desk and type for hours."[10] In 1953, when Kerouac was crashing with his girlfriend Edie (Frankie Edith Parker) and her roommate, Joan Vollmer Adams (the future Joan Burroughs), Kerouac "had become such a speed typist that the carriage bell on his typewriter would ring like an alarm clock. Since he typed all night, it was hard for Edie to sleep."[11] Third, Kerouac made the following declaration: "I'm going to get me a roll of shelf-paper, feed it into the typewriter, and just write it down as fast as I can, exactly like it happened, all in a rush, the hell with these phony architectures and worry about it later."[12]

In Kerouac's desire for the exactitude of reportage there is more than a little of Burroughs's influence. When Kerouac by his own account "officially" began work on *On the Road* in 1948, he was under the sway of a "new philosophy" of Burroughs's called "factualism," where "there is only fact on all levels, and the more one argues, verbalizes, moralizes the less he will see and feel of fact . . . Talk is incompatible with factualism."[13] Not only does Burroughs's factualism require writing rather than "talk," it may even require *type*writing rather than a general writing to succeed. Like the "reports" that Lee writes in Cronenberg's *Naked Lunch*, factualist texts may have no resemblance to anything like an empirical report of lived events, but they are a specific kind of text produced by a specific set of material writing conditions.

Kerouac's desire to produce a kind of proprioceptive writing evokes Charles Olson's "Objectism" as well as Burroughs's factualism. This should not come as a surprise either, because Olson's methodologies are clearly tied to typewriting as well. While all three writers would later work and meet at Black Mountain College, typewriting seems to have provided for a large degree of commonality in their thinking and writing practices much earlier. Alex Albright concurs, noting that "whether Olson influenced Kerouac or vice versa is not clear; most likely, they were influenced in like ways through mutual friends and ideas that were gaining currency after World War II."[14]

Two weeks after Kerouac began writing *On the Road*, his supercharged typewriting assemblage had produced a single-spaced paragraph 120 feet in length, which furled into a scroll three inches thick.[15] There is some disagreement about whether Kerouac had

The 120-foot long manuscript of On the Road.

used architect's paper Scotch-taped together,[16] shelf paper, or Thermo-fax paper[17] for his creation, but everyone agrees that it was long. In a high degree of excitement, he took the manuscript to the office of his editor, Robert Giroux, unfurling it across the floor and shouting, "There's your novel!" Being an editor, Giroux said the last thing that Kerouac wanted to hear: "But Jack, how can you make corrections on a manuscript like that?" Kerouac rolled up his manuscript and left in a huff. In Kerouac's later recollections of the event, however, he does not mention being unwilling to revise the manuscript; his friend John Clellon Holmes actually remembers Kerouac typing a second draft, transferring the contents of the roll onto standard letter-sized paper and editing it in the process.[18] In any event, by dint of his own revisions and those of his editors, Kerouac's unruly scroll was hacked down into a series of discrete paragraphs, which were then strung together into the oldest of fictional prose forms: the picaresque novel.

For Kerouac, the addition of amphetamines and scrolled paper to the typewriting assemblage was the equivalent of adding a turbo-charger and racing cams to a stock engine. Boon notes that Kerouac produced the manuscript of *The Subterraneans* with the same setup in *three nights* in 1953, and that writing *Vanity of Duloz* involved at least the amphetamines and the typewriter (but may not have been typed on a scroll)[19]; Kerouac's own letters show that the manuscript for *The Dharma Bums* was also originally typed on a scroll.[20] This form of typewriting managed to extricate Kerouac from a creative bind by further accentuating characteristics that are already present in touch typing: receptivity to a dictating voice (whether implicit or explicit) and the obviation of the need to check the copy. But despite its modifications of the scene of typewriting, Kerouac's writing assemblage did not do away with limits, and may actually have imposed some new ones. When writing on letter-sized paper, the ending of every page is very specific, but the end of the manuscript itself is indefinite. On the other hand, typing on a scroll (or typing on drugs, for that matter), foretells an impending ending, one that can be forestalled artificially by taking more drugs or taping more paper to the end of the roll, but there will always be an end to the run. Kerouac himself finally found his typing assemblage ultimately too limiting, a "horizontal account" that he wished to abandon in favour of something "vertical, metaphysical."[21] That *On the Road* is generally considered to be Kerouac's best work is eloquent testimony to the value of at least a baseline form of discipline in creative endeavours, even (especially?) when pushed to their limits and run at maximum speed.

Truman Capote's famous dismissal of Kerouac's work – "isn't writing at all, it's typing"[22] – turns out to be entirely accurate. Capote first made this remark on David Susskind's television show during an appearance with Dorothy Parker and Norman Mailer, but, knowing a bon mot when he uttered one, repeated it as often as possible in interviews in later years.[23]

What's odd is that Capote doesn't see his own brand of New Journalism as an equal but different product of typewriting, rooted, as it is, in many of the same values as Burroughs's and Olson's notions of writing as reportage capable of conveying truth. After repeating his Kerouac joke in a later interview, Capote was asked by his interlocutor exactly how many writers were just typing, to which Capote responded, "Ninety-nine-point-nine percent. (*Laughs.*) And that's being generous."[24] At this point Capote was missing the obvious, even though he had already stated it: writing *has become* typewriting, and he the anomaly. This moment of blindness is even stranger when one considers that it occurred when Capote was fully aware of the difficulties Parker and Mailer were having in attempting to cope with another new medium, television, which Capote had already mastered: "Dorothy Parker was scared out of her wits, 'cause this was live television, and she was just afraid to open her mouth, and Norman – I kept tripping him up all the time."[25]

Even after his comeuppance at Capote's hands, Mailer, in an article for *Esquire*, defended Capote on the grounds that he was invoking "the difficulties of the literary craft in contrast to Mr. Kerouac's undisciplined methods of work."[26] He, too, missed the point. Disciplinary practices saturate Kerouac's writing, but as they are not the kind of disciplines familiar to Mailer, Capote, or Parker, they are effectively invisible. What Kerouac did when he typed was of an entirely different order than what the writers with pens did in his own or the previous century. Boon notes that *On the Road*, which celebrates speed as a value in and of itself, is (like custom cars) a product of "the machinic accelerations of World War II"[27] . . . accelerations which were produced before, during, and after the war, with the aid of typewriting.

Chapter 30

Typewriters at War

Ballistics

"You will see how accelerating finger momentum makes possible the fast, easy stroking called *ballistic*," writes the former Lieutenant Commander Dvorak in *Typewriting Behavior*. Likewise, Friedrich Kittler translates Timothy Salthouse's 1984 statement from the German: "'The typing of a given output resembles a flying projectile' because 'it only needs a starting signal' and 'then goes all by itself.'"[1] In the same year, some of my friends (I wasn't in the typing class, and *still* can't touch type) came to the same conclusion in the John Taylor High School typewriting lab, when they discovered that placing a sharpened pencil with its eraser on the H slug of an electric typewriter and then resting its shaft perpendicularly across the platen created a small but dangerous missile launching device.

Even beyond high school typing class, there is an intimate connection between mechanization, speed, and violence. For architect and theorist Paul Virilio, there was no industrial revolution. Instead, he sees the changes that the mechanization of society brought about as *dromocratic*,[2] that is, characterized by compulsive movement. Where Foucault sees systems of knowledge/power,

Virilio sees systems of moving/power. Society, he writes, will divide into "hopeful populations," who are allowed to believe that they will one day be able to fulfil their desires by speeding off into the infinite, and "despairing populations," who, because of the inferior technology available to the poor, must relegate themselves to slow subsistence in a finite world.[3] Class struggle, in other words, is a measure of dynamic efficiency.[4]

This does not mean, on the other hand, that it is enough to be fast in order to escape the various societal systems of discipline; "revolution is movement, but movement is not a revolution."[5] In other words, discipline is perfectly capable of shifting gears and kicking *itself* into high speed, as preparation for war.

The first usable modern typewriters were the product of the famous firearms producer E. Remington & Sons, introduced at a time when the end of the Civil War had dried up government contracts for new guns. If, as Clausewitz's famous dicta suggests, war is the continuation of politics by other means, then when war slackens its pace slightly, the finely honed mechanisms for imposing extremely efficient forms of discipline must find a new target to justify their ongoing existence.

The weapon whose name became synonymous with the typewriter, though it was not manufactured by Remington, was invented by a former Remington Chief Engineer, John Taliaferro Thompson. Before working for Remington, Thompson had directed munitions supply for the U.S. Army during the Spanish Civil War. After this conflict, he was responsible for the testing on cadavers and beef cattle in the slaughterhouses of Chicago that led to the Army's adoption of the AS .45 calibre bullet as "the only acceptable cartridge for a handgun." When he retired from military service in 1914 Thompson went to work for Remington with the goal of perfecting an automatic rifle and was responsible for establishing several rifle factories for the company. Thompson understood guns, and he understood the principles of the assembly line equally well.[6]

Thompson was recalled to active duty for World War One and promoted to Brigadier General, then retired again in 1918 to focus on the creation of an "intermediate" automatic weapon – something between the pistol and the rifle – which he believed would be a key piece of technology for future soldiers, but which wouldn't entail the substantial patent licensing fees accompanying the use of existing technology in the hands of major manufacturers like Remington.

By 1919, with the AS cartridge as ammunition, Thompson had created a hand-held machine gun he nicknamed the "trench broom," presumably for the role he hoped it would play in wars yet to come. Like Remington at the end of the Civil War, though, Armistice forced Thompson to change his plans, and he began to market his gun as a law enforcement tool. He renamed his weapon the "submachine gun." The press, wanting something catchier, called it the "Tommy gun," and Thompson registered that name as well, even stamping it on the barrel of some of his weapons.[7] Police weren't the only ones who found the gun appealing. Criminals of the Prohibition era, when civilians could legally possess automatic weapons, fondly called it the "Chicago typewriter." (By 1976, when the weapon was declared obsolete, about 1,700,000 of them had been produced.[8])

During the same period, the typewriter was firmly entrenching itself in the military as an essential, non-metaphorical element of any battle. "When man goes to war he keeps his typewriter close by his side," writes Bliven. In a passage that would later be heavily cribbed by McLuhan,[9] Bliven goes on to assert, "The captain of a battleship insists that there be fifty-five typewriters on board before he feels fully equipped to meet the enemy" – leaving one to wonder if Lieutenant Commander Dvorak served on that particular battleship – and "On the ground, as the army moves forward, there are more writing machines within four thousand yards of the front than medium and light artillery pieces combined."[10]

The Allies, however, were not the only ones with typewriters in their arsenals.

Typewriters and Fascism

A passage near the opening of Kurt Vonnegut's novel *Mother Night* points to a small but significant component of typewriting's role in the creation and maintenance of fascism in Nazi Germany. The novel's narrator is one Howard W. Campbell, Jr. Like Pound, Campbell is an American, and, like Pound, he joined an Axis power as a radio propagandist during World War Two, broadcasting messages extolling the virtues of fascism and excoriating the policies of the Allies to Allied troops. At the opening of Vonnegut's novel, Campbell is in prison in Jerusalem, awaiting trial for his war crimes.

Campbell has been given a typewriter by one of his captors, Mr. Tuvia Friedmann, director of the Haifa Institute for the Documentation of War Criminals. Campbell begins with the observation that Friedmann "has expressed an eagerness to have any writings I might care to add to his archives of Nazi villainy. He is so eager as to give me a typewriter, free stenographic service, and the use of research assistants, who will run down any facts I may need in order to make my account complete and accurate."[11] In the face of the incomprehensible atrocities of the Holocaust, Friedmann relies, once again, on the abilities of typewriting to produce truth from even the most recalcitrant of writing subjects.

The typewriter that Friedmann gives Campbell to use is not just any typewriter, either.

> It is a curious typewriter Mr. Friedmann has given to me – and an appropriate typewriter, too. It is a typewriter that was obviously made in Germany during the Second World War. How can I tell? Quite simply, for it puts at finger tips a symbol that was never used on a typewriter before the Third

> German Reich, a symbol that will never be used on a typewriter again.
>
> The symbol is the twin lightning strokes used for the dreaded *S.S.*, the *Schutzstaffel*, the most fanatical wing of Nazism.
>
> I used such a typewriter in Germany all through the war. Whenever I had occasion to write of the *Schutzstaffel*, which I did often and with enthusiasm, I never abbreviated it as "*S.S.*," but always struck the typewriter key for the far more frightening and magical twin lightning strokes.[12]

In the twin lightning bolts, terror takes on its own typewritten signifier, hurled onto the page by the touch of a trigger just as ordnance is hurled toward the enemy. Paul Virilio observes that in such a context, everything becomes a question of time, and that this is the origin of symbols such as the twin lightning bolts of the SS. "Speed is Time saved in the most absolute sense of the word, since it becomes human Time directly torn from Death – whence those macabre emblems of decimation worn down through history by the Assault troops, in other words the *rapid troops* (black uniforms and flags, death's heads, by the uhlan, the SS, etc.)."[13]

What Campbell does not mention is that the modern typewriter itself is already a concrete manifestation of speed, and therefore amenable to utilization by the forces of terror; the SS key is merely a superinducement. One of the interesting features of this passage is that it immediately punctures the elaborate fiction of the book as a whole. *Mother Night* begins with a fictitious Editor's Note, in which Vonnegut positions himself as Campbell's editor, insisting that his creation is a real person. But nowhere in the text does the "magical and frightening" twin lightning-bolt symbol that is so particular to Nazi typewriting appear, despite the ease with which modern typesetting could accomplish this. There are only the initials "S.S.", rendered exactly in the manner that Campbell makes a point of stating he did *not* type them. Perhaps there are some truths that cannot be typewritten after all. Despite Friedmann's careful preparations and Vonnegut's elaborate framing device, the assumption in *Mother Night* that the typewriting can produce a true account of Nazism collapses utterly, and not even two full pages into Campbell's narrative.

But if typewriting plays a role in the waging of war and the creation of terror, it can also be used to fight against it.

Samizdat

"Samizdat" is Russian for "self-published." In general use, it refers to a system that was developed in Soviet bloc countries to copy and circulate works of literature, politics, and philosophy that had been officially suppressed by the government.[14]

Before Glasnost, any Soviet business or institution that dealt with scientific or technical information, or had access to printing technologies, had a branch of the First Department on their premises. The First Department, which reported directly to the KGB, controlled and inventoried all printing presses, word processors, photocopiers, and typewriters, ostensibly in the interest of ensuring national security and protecting state secrets.[15] In practice, though, this process resulted in the often violent suppression of free speech, the right to publish, and other basic rights.

Samizdat culture evolved as a direct response to this censorship. It operated by volunteers reproducing texts in microeditions created with the help of any available technology, down to the level of pen and carbon paper. These texts then circulated by hand from person to person, in some cases, spawning further "editions" along the way; under oppressive conditions, copyright becomes just another mechanism of control.

In the former Czechoslovakia, typewriters played a major role in samizdat culture. After the Communist Party of Czechoslovakia seized power in 1948 and the country fell under Soviet influence, the authorities disbanded the Czech Writers Union and set up a state-sanctioned organization, stripping official status – and therefore the right to publish – from the majority of working professional writers (many of whom had also been sanctioned under the Nazi occupation of 1939–45).[16] The writers, however, found a loophole. While publishing unauthorized books was illegal, circulating manuscripts wasn't. Though manuscript literally means "hand-written," by this point, typewriting was so inextricable from the writing process worldwide that under Communist Czech law a "manuscript" could be typed, or even a carbon copy of a typewritten manuscript, as long as the author signed each copy.[17]

The print run of a samizdat edition varied considerably as a result of the hazardous conditions of their production. Some samizdat presses produced thirty to sixty copies,[18] others, "a maximum of 10, or on electric typewriters 15, copies."[19] In almost all cases, the typescripts were hand-bound for the sake of durability. If a title was popular enough, samizdat publishers would sometimes produce multiple editions. Some texts would have up to a thousand typewritten copies in circulation.[20]

This impressive scale of production occurred despite the samizdat publishers' reliance on poorly trained volunteer typists who had to place themselves and their families in danger in order to do the work. The Czech secret police regularly used tactics including home inspections, wiretaps, mail checks, confiscation of identifying documents, beatings, and death threats in their attempts to keep samizdat typists in line.[21] It is under these circumstances that the analogy between typewriters and guns reappears. Czech writer and volunteer samizdat typist Zdena Tomin writes, "The use of the typewriter-and-carbon-copies technique is not some kind of elitist snobbery [presumably, as opposed to the now-prosaic pen and carbon paper – a neat reversal of the previous century's assumptions about the appropriate tools for writing]: the possession of 'unauthorised use' of any kind of duplicator may lead in

Czechoslovakia to a heavier punishment than possession of a gun."[22] While this analogy bears the force of the repressive law that suggests it, it is also what makes the typewriter a potent weapon in the eyes of revolutionaries. Writes Tomin, "my typewriter was but a little Private among Generals."[23]

Despite the possible repercussions for its authors and publishers, Czech samizdat thrived until the restoration of democracy in the region in 1989. It introduced to the world the writing of authors such as Vàclav Havel, Ivan Klíma, and Alexander Kliment, and created a legacy of resistance that has become the stuff of popular culture. As science fiction writer and journalist Bruce Sterling once deadpanned about an agent of the U.S. National Security Agency at a conference on computer privacy, "I've seen scarier secret police agencies than his completely destroyed by a Czech hippie playwright with a manual typewriter."[24]

Electric typewriters, on the other hand, are an altogether different proposition, one that signalled that typewriting was skidding off the royal road for good.

Chapter 31

Electrification and the End of the Grid

Electricity and typewriting have a long mutual history. Pastor Hans Rasmus Johann Malling Hansen was experimenting with electrical carriage movements from as early as 1867, as we've seen, making his writing ball not only one of the first functioning typewriters, but very likely the first electric typewriter. But it wasn't until 1923 that anything resembling a fully electric typewriter appeared. And that machine was the beginning of typewriting's end.

In the early 1920s, an inventor named James Smathers had perfected a design for a practical electric typewriter, a design that was subsequently improved by Russell Thompson, an employee of the North-East Electric Company. After a year of steady development, the North-East Appliances Company began production of Thompson's design. Sales were brisk enough that within five years the company spawned a separate division, Electromatic Typewriters Inc. Four years later, it was absorbed by the International Business Machines Corporation, and became IBM's Electric Typewriter division.

IBM had something that Electromatic didn't: capital to invest in research and development. They ploughed more than a million dollars into their typewriting division, and in 1935 launched the first commercially successful electric typewriter.[1] But the electrification of typewriting was not nearly as significant as what was to come next.

Six years later, in 1941, IBM broke the grid.

For as long as typewriting had existed up to this point, each character on a typed page occupied the same amount of space, implicitly dividing a blank sheet of paper up into an invisible grid, with each cell capable of holding exactly one character, be it an m (the widest character) or a period (the smallest). This invisible grid is the essence of the disciplinary structure that typewriting applies not only to typewritten texts, but also to the bodies of typists and ultimately to larger society, adapting the world to its exigencies. Writers have, by turns, celebrated and excoriated the grid, but ultimately had to either submit to its logic or turn to other means of inscription. Printers remained confident in the belief that movable type was inherently superior to typewriting because of its ability to finely adjust the spacing of letters in a way that the typewriter could not.

The deep coffers of IBM changed all that by allowing their engineers to develop a mechanism that broke each cell of the grid into fifths. In 1944, IBM launched the Executive, a proportionally spaced typewriter. Characters on the Executive typewriter occupied between two and five units per grid cell, depending on the width of the letter. Beeching relates an anecdote that demonstrates the significance of this achievement. The proportionally spaced typewriter immediately leaped to the apex of the world bureaucracy and administrative culture when President Roosevelt was presented with the first machine off the line. The Armistice documents that ended World War Two were typed on an IBM, as was the original United Nations Charter. To a world accustomed to monospaced typewritten documents, a page of typewriting produced with an Executive (and here, finally, the machine name becomes a metonymy for the dictator as well as the amanuensis) was indistinguishable from a page of typeset text. Prime Minister Churchill allegedly responded to Roosevelt that "although he realized their correspondence was very important, there was absolutely no need to have it printed."[2]

In 1961, IBM introduced another radical change to typewriting technology: the "golf ball" typehead of the IBM Selectric. This was a single, spherical element that replaced all eighty-eight typebars. Because the golf ball moved along the page on a metal bar during typing, the typewriter carriage also disappeared, making typewriters more compact, quiet, and efficient. Moreover – and this is the truly radical aspect of golf ball technology – for the first time there was no definite physical connection between the letter key that a typist pressed and the mark that appeared on the paper (in 1952, IBM had built a typewriter with changeable type bars, but they were cumbersome and expensive compared to the golf ball).[3] In effect, swappable golf balls meant that typing one key could produce any conceivable symbol as a result, even a result unknown to the typist until the instant of its inscription. The moment that was unthinkable to the fevered imaginations of the Italian Futurists ("For a typewriter to have its E pressed and to write an X would be nonsensical. A broken key is an attack of violent insanity") had not only come to pass – it had become the standard for all of typewriting. Typewriting had fewer than five years left before its summary execution in the Nevada desert at the hands of Ed Ruscha.

Other things were also happening at IBM – big things. In 1944, IBM manufactured "the automated sequence controlled calculator," a practical large computer. In 1957, they coupled their computers with an "input output" typewriter, which could be used to both convey commands to a computer and to print computer output at twelve characters per second.[4]

There are four points worth observing here. First, though typewriting precedes computing, it has little to do with computing's development or internal logic, and only becomes a convenient interface for a computer thirteen years after the fact. Like the golf ball itself, it can and will be swapped for other kinds of interfaces (punch cards, tape, electronic or acoustic signals, voice, stylus, and tablet). Second, computational speed has little to do with typing speed. When computers begin to connect together, the speed of typed output becomes an irrelevant consideration. Third, both computing and proportional spacing are wartime technologies. They appeared at a time when IBM was also producing automatic rifles and carbines, bomb sights, director and prediction units for anti-aircraft guns, and other kinds of ordnance, accepting only a 1 per cent profit from the U.S. government in return.[5] From the Remington to the IBM, the connection between typewriting and guns was never innocent.

It is only fitting, then, that the last word should go to a writer equally comfortable using either a typewriter or a shotgun to make his art.

Epilogue

The Typewriter in the Garden

A fragment from one of William S. Burroughs's final interviews, conducted over the telephone by Lee Ranaldo of Sonic Youth, approximately four months before Burroughs's death in 1997:

LR: Y'know, I have one last question for you – is that, uh, typewriter still growing out in your garden?

WSB: (*puzzled*) What typewriter?

LR: Last time we were out there to visit, you had a typewriter growing in your garden amongst all the plants and things . . .

WSB: Oh, just threw one away I guess . . .

LR: Yeah, it was a very beautiful image there, with the weeds coming up through the keys . . .

WSB: (*laughs*) I guess so – I don't remember the typewriter – I've gone through so many typewriters – wear 'em out and throw 'em away.

LR: Do you generally write with a computer these days?

WSB: I have no idea how to do it. No, I don't.

LR: Typewriter or longhand?

WSB: Typewriter or longhand, yes. These modern inventions! James [Grauerholz, Burroughs's companion] has one, but I just don't.

LR: Okay, well listen William, I thank you very much. Please tell both Jim and James thanks for their help as well.

WSB: I certainly will.

LR: Okay, you take care.

WSB: You too.

LR: Bye bye.

WSB: Bye bye.[1]

Part

Aftermath: Typewriting After the Typewriter

They say the computer is an improved form of the typewriter. Not a bit of it. I collude with my typewriter, but the relationship is otherwise clear . . . I know it is a machine; it knows it is a machine.

– Jean Baudrillard

Building the Chimera

After typewriting is shattered into a million pieces, does any of its logic persist into computing? Laptops look like late-model typewriters . . . *sort of* . . . but appearances are just that: appearances. Like everything else mediated by computing, typewriting has entered the world of simulation: the image now precedes the actuality.

The retrofuturist computers in Terry Gilliam's *Brazil*, apparently cobbled together from parts scavenged from both the nineteenth and twenty-first centuries, are no longer simply images. At least, not in parts of the computer case modifying ("casemodding") community. Casemodders have inherited the tradition begun by car and motorcycle fanciers, chopping and styling their plain-vanilla computers into a staggering array of unique machines, and displaying them all over the web.[1] It was only a matter of time before it occurred to someone to turn Gilliam's part-nostalgic, part-ambivalent aesthetic into a blueprint for a working machine.

That someone turned out to be Joel Zahn, a computer technician and antique typewriter collector from British Columbia. In his illustrated step-by-step description of the mod, Zahn writes, "Quite a while back, one of my friends sent me a 'photoshopped' picture of an old typewriter (it looked like a Remington) fitted with a monitor, mouse, and printer (the printer had the platten [*sic*] and carriage attached to it.)" After wangling with a number of the used typewriter relic-mongers on eBay and purchasing several machines that turned out to be in better condition than the machines already in his possession, he ultimately decided it would be less trouble to dismantle the first antique typewriter he had purchased, an Underwood No. 5, manufactured in 1924, and use it as the frame for his mod.[2]

A month's worth of work in his spare time produced a working computer housed entirely inside the old Underwood frame. To complete the simulation, Zahn installed a free program that allowed him to map sampled typewriter sounds from one of his original Underwood typewriters onto the electronic keyboard, making his "digital Underwood" sound like the original.[3] Which begs the question: does the word "original" have any place in this context?

The Death of the Typing Classes

The keyboard itself has never been the only interface for computing; a few of the more common ones include touchscreens, mice, stylus-based systems, simple keypads, cards and magnetic strips, direct network input, and voice recognition. As the satirical newspaper *The Onion* suggests, the keyboard is in no particular danger at the moment:

> NEW YORK – Fidelity Financial Services' Gwen Watson, 33, shouted angrily at her IBM ViaVoice Pro USB voice-recognition software, sources close to the human-resources administrator reported Monday. "No, not Gary Friedman! Barry Friedman, you stupid computer. BARRY!" Watson was heard to scream from her cubicle. "Jesus Christ, I could've typed it in a hundredth of the time." After another minute of yelling, Watson was further incensed upon looking at her

> screen, which read, "Barely Freedman you God ram plucking pizza ship."[4]

Humour aside, this does not mean that the keyboard will never wither away and disappear, as Marshall Jon Fisher suggests in "Memoria Ex Machina":

> Perhaps . . . this very act of typing is what will linger one day in my mind's reliquary. Voice-recognition software is pounding at the gates; video mail seems every day more feasible. How much longer will our computers even have keyboards? Typing may someday survive only as another sense memory. A writer, while composing with his voice, will still tap his fingers on the desk like an amputee scratching a wooden leg. Rather than the ghost of a particular machine, it will be this metacarpal tap dance, an apparition of the way we used to express language, that will haunt him.[5]

The notion of a digital competence that extends beyond and even trumps the keyboard is already finding its way into the public school system. Stanley Johnson, director of instructional technology for the District of Columbia Public Schools, says, "We've seen recently the proliferation of cellphones, digital cameras [and] PDAs," which he argues are "just as powerful" as the written word: "Digital literacy skills need to be introduced as well."[6]

The *Christian Science Monitor* cites a U.S. Department of Education report illustrating that the number of public school students in typing or keyboarding classes is at an all-time low. In some schools, typing classes have not been offered for decades, even though the rise in computing technologies would suggest that typing skills would be of increasing relevance. As I've suggested, computing brings with it an arsenal of tools that mitigate the need for speedy data entry, especially cut and paste, in combination with massive storage and searchable archiving capabilities. After consulting with a local staffing agency, the *Monitor* also reports that as recently as the early 1980s, a professional administrative assistant would have needed a minimum typing speed of 50 words per minute; a contemporary administrative assistant might only require a speed of 30 words per minute. Contemporary employers are more interested in a prospective employee's familiarity with basic software packages and ability to format complex documents.[7]

On Slashdot, the web-based newsforum that constitutes the shiny silicon heart of the open-source geek universe, the *Christian Science Monitor* article was debated at length (as is almost any topic that finds its way onto Slashdot). The more insightful comments argued that the requirements of software development have little to do with typing speed. One user, forgetmenot, writes that "how fast you can type code has absolutely no relation that I can possibly think of to effective coding because good code is generally code that was well thought out and designed prior to 'typing' the first line. Typing faster without thinking about the design just means you make design mistakes all that much sooner. Furthermore, the keystrokes in a typical program usually resemble nothing like prose, so learning to type probably doesn't help much."[8]

The Space Between

The computer is not a typewriter: philosophers and typographers, two of the unlikely kinds of authorities that dominate the arguments in this book, agree on this point. But *why* and *how* is computing different?

In just a handful of years, cultures sometimes stop thinking in the way they have been and begin to entertain new ideas in an entirely different way. Foucault calls this space that appears between modes of thinking (and the technologies that help to create and propagate them) a *discontinuity*. A discontinuity "probably begins with an erosion from outside" – an erosion caused by something external to the dominant order, "but in which it has never ceased to think from the very beginning."[9]

The idea of automated writing not only precedes typewriting, it aids and abets in its creation, and accompanies it as a kind of counterpoint throughout its existence. It led to the creation of the writing automata, which donated principles leading to the construction of the first typewriters. It is the logic that leads to mechanically inscribing a scroll, particularly a punched scroll that can be replayed to produce or reproduce the effects that created it. This is the logic behind Peter Mitterhofer's first machine, which used pins to perforate letters on the page, and to an extent, it is present in the thinking of Olson and others who saw the typewriter as the mediator in the process of accurately capturing a writer's thoughts and moods.

Nevertheless, the patterns of knowledge that describe computing and the patterns that describe typewriting are entirely different. The opening of Robin Williams's *The Mac Is Not a Typewriter*, one of the most popular computer books ever, details how these differences begin on the most minute and highly structured visible level – the ordering of letters on the page – and expand outward to encompass the ordering principles on which bodies, corporations, and societies are constructed.

Williams begins with the observation that the habit many typewriter-trained typists have of hitting two spaces after a period is completely unnecessary when using a word processor. Typewritten characters are monospaced, so the period, the smallest character, occupies the same amount of space as the m, the widest character. Virtually all computer typefaces, however, are proportionally spaced; each character only takes up as much space as it needs. Most characters occupy a fifth of the space that the letter m requires.[10] With proportional spacing, a double space after a period leads to gappy channels of white space running down the page, an unsightly effect that typographers call "rivering." A touch typist who has conditioned herself to double-space after periods discovers that when using a word processor, this former asset has become a liability.

The consequences of hewing to the old order from within the new one are not simply aesthetic; they're also ontological. Williams tells the reader that continuing to use two spaces after a period "creates a disturbing gap,"[11] perhaps because it mirrors that gap between the old logic and the new one. Bridging that gap is not particularly easy, because it evidently requires not only discipline, but also the use of admonishings and ridicule as incentive: "Your eye does *not* need that extra space to tell you when the next sentence begins"; "In unpublished papers, email, personal correspondence, use as much space as you feel comfortable with. But once it gets to press, don't make yourself look foolish."[12]

The system of proportional spacing itself allocates priority according to different criteria than the system of tables that characterize the typewriting grid. Where the typewriter attempts to standardize bodies and spaces to function according to its logic, computing is more flexible and adaptable, wrapping closely around and even insinuating itself into bodies and spaces to the extent that a clear distinction between biology and technology becomes problematic – hence the emergence of the cyborg as the mascot of postmodernity.

Perhaps the cyborg – an uncertain and unpredictable fusion of human and computer – holds the possibility for liberatory and transgressive gestures,[13] but in the descriptions of the material conditions of writing in some of the most interesting contemporary texts, writing cyborgs are, if anything, more abject than typing cockroaches. The following passage, from Jack Womack's novel *Ambient*, describes the fate of the descendants of the Type-Writer Girl:

> Each processor sat in a small cubicle, their eyes focusing on the CRTs hanging on the walls before them; each wore headphones so as to hear their terminals – number eights – as they

> punched away. A red light flashed over one of the cubicles. One of the office maintenants rolled over and unlocked the stocks that held the woman's feet. It guided her across the room, toward the lav; her white cane helped her in tapping out the way. The system had flaws; some employees went insane – they were fired – and some grew blind – the ones whose fingers slipped were given Braille keyboards, at cost.[14]

In *Ambient* the cognate of proportional spacing is the ability of "the system" to wring every last drop of productivity out of a human asset – the weakest component in the new human-computer writing network – by adapting itself to the steadily degenerating bodies of its employees. The cost for the necessary adaptations, which are already minimal, thanks to the power and flexibility of computing technologies, can always be passed on to the workers themselves.

The situation for generative typists is not much better. The familiar dictating voices are still present, but in a networked milieu, they have become even more despotic, as this fragment of a sentence from William T. Vollmann's *You Bright and Risen Angels: A Cartoon*, demonstrates:

> The keys of my typewriter depress themselves and clack madly, like those of a player piano, like (more appropriately still, since we are in the age of electricity) a teletype machine in some computer center at three in the morning, with the lights glaring steadily down, failed programs in the wastebasket and punchcards on the floor; and far off somewhere at the other end of the dedicated synchronous modem line, a sunken computer swims in its cold lubricants and runs things, and there is nothing to do but wait until it has had its say; the keys do not feel my touch; they do not recognize me; and all across the room the other programmers rest their heads in their arms as Big George dictates to them as well,

> garbage in and garbage out, screwing up everything with his little spots of fun, refusing to drown in the spurious closure of a third-person narrative (think how lonely he must be if he has to play such stupid games with me); when what I really wanted to do was write about our hero . . .[15]

As recently as 1967, the focal character of John Barth's *Lost in the Funhouse* was still capable of formulating elaborate fantasies of authorial sovereignty, describing writing as "a truly astonishing funhouse, incredibly complex yet utterly controlled from a great central switchboard like the console of a pipe organ," and himself as its secret operator.[16] *You Bright and Risen Angels* abandons any hope of mastery; it reveals the fantasy of authorial control as a shimmering chimerical product the funhouse mirrors. The author is out of control from the beginning, merely a local node soldered into the complex network that constitutes the scene of computerized writing. There is no certain point of origin for the text, and, it suggests, no privileged final version. A vast, impersonal, remote mainframe and the villainous Big George dictate simultaneously to the author, who situates himself as one of a masochistic group of "programmers" who only experience subjectivity intermittently: "All I can hope to do is to type in a little ameliorating detail here and there so that my angels will at least have the dignity of consistency as they are made to kill each other, and fall and die, and maybe Big George will draw a long breath at the end of this section and I can make adjustments, but I doubt it, I really doubt it; and all I can say is that I'm very sorry and that I'm dying, too."[17]

The Abstract Typewriter

The discontinuity between typewriting and computing is more than a matter of spacing. As much as anything else, it is due to the application of the chief technology that the typewriter deploys to create discipline – the table or grid – upon itself.

In 1936, Alan Turing had an idea for an imaginary machine. It would be "universal," in the sense that it could simulate the functions of any machine whose primary purpose was the creation or manipulation of symbols or inscriptions. In *War in the Age of Intelligent Machines*, the example Manuel De Landa uses to illustrate the principles of the Turing Machine (as it is now usually referred to) is the typewriter:

> Turing realized that the internal workings of typewriters, calculators and other physical contraptions like them could be completely specified by a "table of behavior." A typewriter, for instance, may be depicted as consisting of several components; the keys, the typing-point, the upper and lowercase lever and so on. For every combination of these components, the machine performs one and only one action: if the machine is in lowercase, and the letter key "A" is pressed, and the typing-point is at the beginning of the page, the machine will print a lowercase "a" at that position. If we were to write all the possible combinations and the resulting machine actions, we would abstract the operations of the machine as a list. By looking up the entry for any particular combination of components (lowercase, "a", start page), we could tell exactly what the machine would do. If we then built our own machine that could read the list of combinations and perform whatever action the list indicated as appropriate, we would be able to simulate the workings of a typewriter. In a very definite sense, the list or table of behavior would be an "abstract typewriter." Similarly, for other machines, we could assemble appropriate lists of behavior and then carry out that behavior with our new device.[18]

Turing Machines can even simulate other Turing Machines. De Landa points out that our contemporary computers are actually a series of several nested simulations: a Turing Machine burnt into a microprocessor simulates another Turing Machine in the form of the computer's programming language, which in turn is used to produce an even higher level of Turing Machine in the form of applications like word processors.[19]

This ability to simulate both a writing machine and the data that it creates or controls is implied in the famous McLuhanism that "the 'content' of any medium is always another medium."[20] However often this phrase is repeated, its full meaning is rarely discussed. If a machine exists as data alongside the data that it produces, it too can be modified and manipulated, even while it is in the process of operating. The abstract machine itself can be the force that is operating on itself; for that matter, so can that data on which it operates. A mechanical calculating device, no matter how sophisticated, cannot modify itself. All that's necessary to make this possible is to provide a common storage space for the abstract machine and its data. John Von Neumann, one of the scientists that constructed the ENIAC computer, the world's first electronic digital computer, during World War Two, referred to this common storage space as the "one organ"; for us, this space is the storage disk, but it doesn't have to be anything so elaborate. In Turing's imagination, the "one organ" took the form of the simplest possible system: an infinitely long paper tape.[21]

While there is a surface resemblance between the "one organ" as a paper tape and the scrolls of late typewriting, and while the archive of computing overlaps with the archive of the typewriter at some points, the bulk of computing's past and future lies elsewhere. One trajectory passes through the player piano rolls invented by Henri Fourneaux in 1863,[22] and, before them, Charles Babbage's 1834 Analytical Engine and 1821 Difference Engine,[23] and, before that, the punch cards that Joseph Jacquard invented to program his loom in 1805. The archaeology of computing is partly the archaeology of software.

Another line runs through the first long-distance transmission of alternating current in 1891 near Telluride, Colorado, followed shortly by a similar experiment in Germany[24]; Samuel Morse's public demonstrations of the telegraph in 1838; Antonio Meiucci's invention of the telephone in 1849, and Alexander Graham Bell's duplication of the feat in 1876[25]; J. R. Licklider's 1962 "Galactic Network" memo describing the principles that would lead to the construction of, first, ARPANET and, eventually, the Internet; Paul Baran's work at Rand between 1959 and 1962 leading to the notion of both a decentralized network and packet switching (paralleled by the work of Donald Davies at National Physical Laboratory, who coined the term "packet switching") and the connection of the first two computers in ARPANET (housed at UCLA and Stanford University) in 1969.[26] The archaeology of computing is also the archaeology of networking.

This is where I part company with Friedrich Kittler's theories on typewriting once and for all. Even if the typewriter is one of the most important devices that can be emulated in a Turing Machine, it is only one such device among many. Further, it doesn't follow that just because computing can simulate typewriting that typewriting produces computing, as Kittler suggests. Computing is neither typewriting's goal nor its conclusion.

As the intermediary between the typewriter and the computer, Kittler proposes the "secret typewriter" – a coding machine invented by German engineer Arthur Scheribus utilizing the principles that would later lie at the heart of Enigma, the cryptographic machine that kept Axis military transmissions safely encoded for most of World War Two, and spurred Alan Turning and the other scientists at Bletchley Park to create the modern computer in an attempt to crack Enigma's codes. Several difficulties with this thesis appear immediately.

The first is that the abstract machine – or what Manuel De Landa calls the "engineering diagram," that is, the underlying principles determining its operation – of the typewriter and of Scheribus's machine are totally different. The "iron whims" of the typewriter are fixed and mechanical, those of the coding machine, electrical and movable. The most widely used principle in twentieth-century coding machines is what David Kahn's definitive text on the subject, *The Codebreakers*, calls "the wired codewheel, the rotor."[27] Each rotor consists of a rubber disk studded with twenty-six evenly spaced metal contacts, connected randomly by electricity-conducting wires to a contact on the opposite face of the wheel. The contacts on the input side of the face represent what cryptographers call "plaintext," the normal order of an alphabet before enciphering. The contacts on the other side represent the "ciphertext," the encrypted alphabet to which the plaintext will be translated. When an input device sends a current through the rotor, it encrypts each letter of the plaintext in turn and sends a signal to an output device.[28]

What makes the rotor interesting as an encoding device is that turning the rotor changes the ciphertext alphabet. To use an example from the indignant Italian Futurists, if pressing an E produced an X before the rotor turned, after one click of the wheel, pressing an E might then produce a Y. All the other plaintext letters would also have different ciphertext letters. Placing all of the possible resulting ciphertext alphabets on a table produces 26 possible ciphertext alphabets. On its own, this is not a remarkably difficult cryptographic challenge, but adding a second rotor between input and output creates 26 × 26, or 676 possible ciphertext alphabets. Five rotors yields 11,881,376 combinations. In Enigma machines, though, each rotor turns one space *each time a letter is typed*. The result is what Kahn giddily refers to as an "outpouring of cipher alphabets in such hemorrhaging profusion as to provide a different alphabet for each letter in a plaintext longer by far than the complete works of Shakespeare, *War and Peace*, the *Iliad*, the *Odyssey*, *Don Quixote*, the *Canterbury Tales*, and *Paradise Lost* all put together"[29]: literature is humbled before cryptography, and a human lifetime no longer long enough to investigate manually each possibility. As revolutionary as this machine is, it has nothing to do with typewriting other than the provisional use of a keyboard as a convenient input device.

Kittler's genealogy of this device, which places its originary moment with his countryman Scheribus in 1919, is also suspect. According to Kahn, Scheribus was neither the first nor the only inventor of the rotor. Like typewriting, the codewheel rotor was independently invented in several different places at several different times. "Spurred by the vast wartime use of secret communications," writes Kahn, "and beckoned by the new age of mechanization, [four men] independently created the machine whose principle is perhaps the most widely used in cryptography today."[30] The first machine that put the engineering diagram of the rotor into practice was drafted by Edward Hugh Hebern, an American, in 1917, and constructed in 1918, a year before Scheribus.[31] A third man, a resident of the Netherlands by the name of Hugo Alexander Koch, filed a patent for another rotor-based "secret writing machine" in Holland

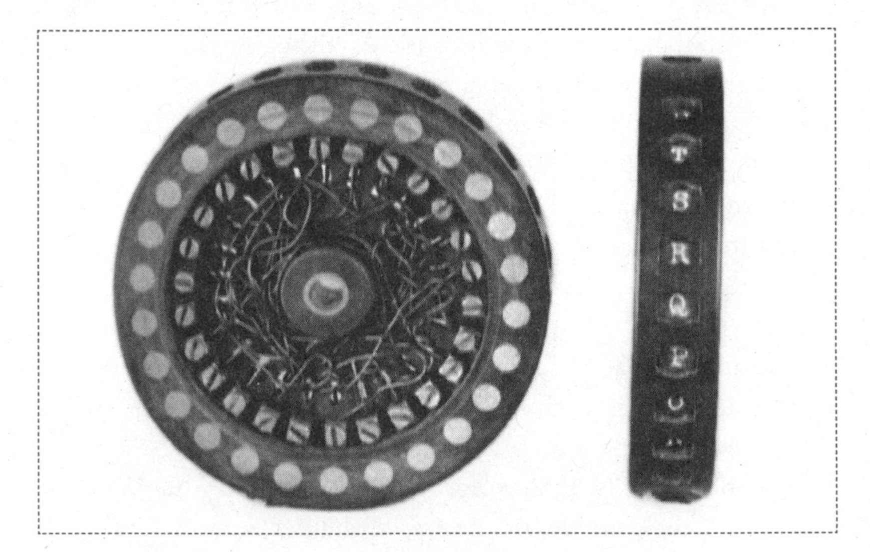

A rotor from the Hebern machine: this is not a typewriter.

in 1919 (in 1927, he assigned his patent rights to Scheribus). The fourth, Arvid Gerhard Damm, a Swede, filed his patent for a two-rotor device three days after Koch, but in Stockholm. Ultimately, Damm is more important for having founded the only commercially successful cipher machine-manufacturing company rather than for his rotor system, which was clumsy in comparison to the other three. Kahn states that Koch viewed the rotor "most comprehensively"; Koch observed that the engineering diagram embodied in the rotor mechanism could also be manifest in a system made of any of the following: steel wires on pulleys, rays of light, or fluids in tubes. Moreover, the coding impulse could move through a system of bars and plates rather than rotors[32] . . . demonstrating, once again, that the semblance of encoding machines and typewriters is purely cosmetic, and has nothing to do either with their basic diagrams or the kinds of discourses that they create.

There are surface resemblances between typewriting and computing because the QWERTY keyboard has become our default *interface* for computing, but computing is a discourse whose rules are determined by the functioning of software and networks, not by mechanical devices and hierarchies. Computing insinuated itself into typewriting as it did into everything else, but the hybrid electronic typewriter/word processors of the late 1980s were a retroactive attempt to graft aspects of computing onto typewriting, not a sign of typewriting in the process of becoming computing. Neither one thing nor the other, they are consigned to the roadside trash even more quickly than antique typewriters, because they have no nostalgic value, on the one hand, and cannot substitute for a cheap computer, on the other. Attempting to push typewriting beyond the limitations of a mechanical system with limited potential for alteration through feedback and extension brings the entire discursive formation to a screeching halt, and typewriting shatters itself across the desert of the Real.

In the meantime, Kittler writes, "By consciously merging with the writing of world history we comprehend its spirit, we become equal to it, and – without ceasing to be written – we yet understand ourselves as writing subjects. That is how we outruse the ruse of world history – namely, by writing while it writes us."[33] If so, this state has not come about because "world history comes to a close as a global typewriters' organization,"[34] but because of the emergence of bazaar-style global software development networks whose organizational principles are nothing like those of systems organized around the logic of the profoundly non-networkable typewriter.[35] Typewriting reaches the end of the road several decades before Kittler's software-enabled end of history, even if computing's simulation of typewriting to some extent occludes this end. The terms "typewriter" and "typewriting" persist in a networked milieu, but after the discontinuity, they have taken on different meanings and connotations, are arranged, ranked, and ordered differently, and circulate in different contexts, under the control of entirely different discursive formations.

The Future Doesn't Like Us

Gonzo journalist Spider Jerusalem, the antihero of Warren Ellis's graphic science fiction serial *Transmetropolitan*, has an obsession with the Truth. "If the Truth is there," writes Jerusalem, "I will scale the foulest jeweled stairways, eat the vilest gourmet food, and defile my eyes and thoughts with the most beautiful genetically-sculpted prostitutes in order to reach it."[36] A few centuries from now, Jerusalem inhabits, and writes obsessively about, the City, a chaotic dystopian megalopolis of the sort that typifies cyberpunk science fiction.

To paraphrase Matt Groening, in the City, one minute's creepy psychopathology becomes the next minute's wholesome trendy middle-class youth culture affectation.[37] In this perpetually disrupted and disruptive environment, discursive formations come and go in the blink of an eye. Signs persist, but their meaning is never certain because of the underlying changes in the rules that organize the objects and people that comprise the City. One of the running themes of *Transmetropolitan* is that to a citizen of the twenty-first century, this environment of constantly shifting discourse networks would be incomprehensible to the point of inducing insanity.

A Jerusalem column titled "Another Cold Morning" tells the story of Mary, a late twentieth-/early twenty-first-century photographer who'd been cryogenically preserved at death and then "Revived" in Jerusalem's own era. Mary is initially shocked when she's told how long she'd been in suspension, but takes a moment's comfort when a bored and jaded Revival counsellor directs her to a vehicle that's waiting for her outside the City Reclamation facilities, double-parked. "'Double-parked.' She clung to that. It *meant* something, after all; cars, driving, roads, something dully normal. Something real at last," writes Jerusalem, who then adds, "It didn't occur to her that that meant she'd have to go out onto the street."[38]

What follows is a two-page collage of fragmented images of the City, with Mary's staring, incomprehending eyes at the centre, followed by a frame of her lying in a fetal crouch on the sidewalk.[39] This collage is a visual metaphor for the unrepresentable; a contemporary reader can read it quite easily *because* it only stands in for a way of organizing the world that does not yet exist. Its intent is to invoke a profusion of subjects and objects arranged with a logic that would be incomprehensible even to a sophisticated viewer like Mary, who, after all, had professionally produced representations of her own time, and would have been acquainted with all of the vagaries of the digital image. The persistence of the sign "double-parked" in Jerusalem's era is no guarantor that the discursive formation in which it exists will bear the slightest resemblance to the "dully normal" one of cars, driving, and roads that produced it. As Gene Wolfe notes in his Appendix to *The Shadow of the Torturer*, a chronicle of the far future "originally composed in a tongue that has not yet achieved existence" and translated into English, chroniclers of the "posthistorical world" are frequently "forced to replace yet undiscovered concepts by their closest twentieth-century equivalents." If all signs are suggestive rather than definitive, under these conditions, there is an even greater degree of uncertainty about reference.[40]

It is Jerusalem's self-imposed quixotic task to try to locate the statements of Truth mired in this chaotic system on behalf of a readership he despises and loves by turns (as much as he resembles anyone else, Spider could be a tattooed Foucault). Jerusalem is an exemplary Kittlerian subject, written on by the City even as he writes about it. His body is covered in neotribal tattoos, acquired as visible signs of the stories he has written about the City's denizens. Though their individual signification may include falsity and forgery ("I was tattooed with a false entry badge in order to sneak into an Antique Primitive terrorist meeting"), cumulatively, they present a statement about Jerusalem's commitment and ability to find Truth, as most of them represent honours for completed stories ("I was tattooed in Britain before Parliament in Stonehenge for my series on the Catholic War").[41] Similarly, his prodigious appetite for drugs and mass media of every sort, born of the attempt to understand the City by immersing himself in its flows (he writes, tongue lodged firmly in cheek, "I am going to watch City television all day, in order to bring back shining insights about our lives"[42]), writes on his mind as well as his body, eventually producing "information-related progressive mental damage."[43] By the end of the series, an exhausted and physically debilitated Jerusalem, who has, like Hunter S. Thompson and Zarathustra, retired to the top of the nearby mountain (where he had exiled himself at the series' beginning), tells his editor, "Couldn't type if I tried."[44]

How better to indicate Spider Jerusalem's commitment to producing the Truth than with the visual language of typewriting? When Spider is working on a column, the text of his report appears above the illustrations in a monospaced, Courier typewriter-style font. Frequently, as in "Another Cold Morning," the background of Spider's "typed" text box is canary yellow, evocative of the flimsy newsprint sheets utilized in newsrooms to save on the expense of white bond paper. Jerusalem and other characters even refer to his writing machine as a "typewriter."[45] It bears a strong visual resemblance to a typewriter: it tapers from back to front, has a knob on the side where a typewriter's paper scrolling knob would be,[46] and has round keys like those on many antique typewriters. On some of the serial's covers, where, in keeping with contemporary comic book traditions, illustrated objects tend to be more elaborate than they are within the pages of the actual comics, the resemblance between Jerusalem's writing machine and a typewriter is even more pronounced.[47]

Spider Jerusalem, typing the Truth.

Sometimes,[48] but not always, Jerusalem's machine features a QWERTY keyboard. In "Another Cold Morning," as if to counterpoint the degree to which discursive formations and social ordering principles have changed between Mary's time (our time) and Spider's time, his keyboard is radically different: the keys are arranged in diagonal rows and include some icons (a skull and crossbones, an atom) as well as letters, and a lone "function" key (F7).[49] It may be that to some extent, the changing configuration of the keys is just bad continuity or possibly simplified illustration to accommodate a smaller frame, as when configurations change within several pages of each other, during a sequence that depicts one continuous act of typing.[50] But it also shows the instability of the signs "typewriter" and "typing" in Jerusalem's world. As in Gene Wolfe's writing, they are the closest twentieth-century cognates for a machine and process that have yet to be invented. The machine's laptop-like appearance (it folds, with the screen doubling as a lid) and general computer-like capabilities (persistent wireless networking, ability to act as a two-way videoconferencing system) give it some semblance to contemporary computers, but these may be just as tenuous as the connection to typewriters. Jerusalem lives in a time where the planet Mercury has been covered in solar panels to provide power for Earth[51] and people can be downloaded into immortal "foglet" clouds of nanomachines.[52] The specific technology that's actually inside Jerusalem's writing machine, the principles of its operation, and the rules under which it functions, are anyone's guess. Despite its appearance, Spider's "typewriter" could be full of nanomachines.

Like some twentieth-century typists, Jerusalem compares his typewriter to a gun, but for opposite reasons. The description of the Thompson machine gun as the "Chicago Typewriter" or the Kalashnikov machine gun as "the typewriter of the illiterate" emphasizes the power and efficacy of the gun. In his interview with Geert Lovink, filmmaker János Sugár, who made a documentary about the Kalashnikov, says "a gun is a possible direct communication accessory."[53] Even here, the use of "communication" is moot because the message (bullet) destroys its intended receiver, making communication (the transmission of information between sender and receiver) impossible. Conversely, Jerusalem compares his typewriter to a gun to suggest the limited power and capriciousness of the writing machine, and to abrogate his sovereignty as a writer. He says, "I can't control *anything* with this typewriter. All this *is*, is a *gun*."[54] Jerusalem's gun only has "one bullet in it" and is literally hit or miss: "Aim it right, and you can blow a kneecap off the world . . ."[55] but the vagaries of sending reports off into the chaos of a global news network mean that just as often as Jerusalem's writing scores a direct hit, it misses entirely, or only scores a flesh wound. Despite the fact that throughout *Transmetropolitan*, Jerusalem continually scores "direct hits" on his targets with his writing, telling the Truth when no one else can or will, he is usually unable to make a lasting impact. So much for the might of "typewriting."

It's interesting to note that the second time Jerusalem uses the gun metaphor, he says "Journalism is just a gun,"[56] not the typewriter itself. By a process of substitution, "Typewriter" becomes a synecdoche for "journalism," because both connote the possibility of the production of true writing. This remains only a possibility, however, because, as Jerusalem complains, "nobody *does* investigative journalism anymore."[57] In any event, the specificities of the typewriter as a machine are lost somewhere in the distant past.

All that the reader of *Transmetropolitan* has for certain is a visual sign chosen by a writer-artist team as a rough analogue for a future technology, on the basis of nostalgia for a past technology used by journalists in that mythical time when typewriting supposedly produced Truth. As with all good science fiction, *Transmetropolitan* actually turns out to be a reflection of the concerns of the moment (the cover of the final collection, *One More Time*, features a smirking Jerusalem sitting on an hourglass-shaped commode, and pushing a key on his laptop/"typewriter" as a mushroom cloud rises between a building labelled "Bushco" and another bearing the infamous "Mission Accomplished" banner from George W. Bush's aircraft carrier press conference). By the time the era that Jerusalem is supposed to inhabit finally arrives, people may well be nostalgic for the era when people were nostalgic for typewriting, even if all signs of typing have fallen away from acts of composition and communication.

Typewritergate

How far has contemporary culture moved away from typewriting? We can no longer even recognize it when we see it. Or *can't* see it, as the case may be.

On September 20, 2004, something unusual happened: respected CBS news anchor Dan Rather issued a public apology for making a mistake[58] – a *big* mistake. That mistake was largely due to the inability of contemporary experts either to properly verify or definitively disprove the authenticity of a number of typewritten documents, and perhaps even to determine whether they were produced with a typewriter or with a word processor. Martin Tytell, where are you now?

The documents in question were of crucial importance, because they "purported to show that President George W. Bush received preferential treatment during his years in the Texas Air National Guard."[59] As they surfaced during a particularly sensitive phase of the 2004 United States presidential election, when candidates Bush and Kerry were running neck-and-neck in the polls, they had the potential to tip the election definitively in Kerry's favour, not only if they were authentic, but even if they created sufficient doubt about the President's service record.

The four memos obtained by CBS were allegedly written by Lieutenant Colonel Jerry Killian, who had been (Lieutenant) Bush's National Guard squad commander (Killian died in 1984). CBS obtained the documents from Bill Burkett, a former Texas Air Guard commander. Burkett did not approach CBS; the network went to him and asked him for the documents. CBS notes that "Burkett is well known in National Guard circles for a long battle over his medical benefits, and for trying for several years now to discredit President Bush's military service record." While Burkett first asserted that he had obtained the documents in question from another guardsman, he later admitted lying to CBS and changed his story, citing a source the network was unable to verify. Nevertheless, Burkett maintains that the documents are genuine.[60]

Because "typewriter" always has a double signification, referring both to a machine and its operator, what CBS did next should seem obvious: they interrogated the typewriter. Marian Carr Knox, now eighty-six years of age, was Killian's secretary during much of the 1970s. Her testimony to CBS is extremely interesting: "Knox says she didn't type these memos, but she says she did type ones that contained the same information. 'I know that I didn't type them,' says Knox. 'However, the information in those is correct.'" She added, "'It seems that somebody did see those memos, and then tried to reproduce and maybe changed them enough so that he wouldn't get in trouble over it.'"[61] Is the testimony of a typewriter enough in the face of the lack of any actual typewriting? Is "like" as good as "is" in the eyes of today's investigative journalism? How can intent be attributed to a forgery, let alone an "authentic" document? Evidently, the cultural association of truth with typewriting is still so strong that some of the aura of truth persists even in the lack of all verifiable material evidence. If typewriting functioned according to a logic of mechanical production, then, as Baudrillard has famously noted, the dialogue between Rather and Knox would have operated under an entirely different discourse, determined by the conditions of simulation: "It is no longer a question of imitation, nor of reduplication, nor even of parody. It is rather a question of substituting signs of the real for the real itself." In other words, I didn't type these documents, but they look a lot like the ones I did type, and the information they contain is correct as I remember it. "To simulate," Baudrillard states, "is to feign to have what one hasn't."[62]

While the dubious provenance of the documents is ultimately what discredited them, the first clues that there might be something wrong appeared in a literally literal form: the characters on the paper. Moreover, these clues appeared in the context of a new computer networking technology that is challenging to the control traditional journalistic institutions like news networks have maintained until now over the production of truth: weblogging.

Weblogs – a loose term at best, but for our purposes, database-driven web-based publishing systems with built-in syndication, content tracking, commentary, and moderation features – surpass predecessors such as zines and garden-variety web pages by combining the cost-effectiveness of zines with the global reach of the web. Not only are the most popular weblogging software and services free,[63] weblogs remove many of the technical barriers associated with online publishing by automating the process.

Unlike print-based publishing, which adds value to writing by filtering the raw material (editing), and then by investing in the material production of an object (book, article, photograph, film), weblogs serve up great gushing streams of unmediated digital content to anyone who cares to search and browse it. Technopundit Clay Shirky claims that "weblogs mark a radical break. They are such an efficient tool for distributing the written word that they make publishing a financially worthless activity."[64] Weblogging, aided and abetted by computing, is a technology of networks; journalism, synecdocally connected with typewriting, is a technology of hierarchies. Weblogging may not be able to guarantee truth, but, by virtue of its ability to connect, aggregate, rank, and display vast numbers of differing opinions without needing to smooth them over into one definitive narrative, as journalism does, it can definitely create reasonable doubt.

LIBERTY WAITS
ON YOUR FINGERS –
KEEP ON BLOGGING!
http://eff.org/bloggers/

In a world of weblogs, typewriting still signifies Truth . . . even if we don't really know what it looks like any more.

One of the most heavily connected nodes in the net of weblogs asking questions about the CBS story was and continues to be a right-wing weblog called Powerline. In a post, and long subsequent thread of discussion, titled "The Sixty-First Minute,"[65] which appeared on September 9, 2004, the site administrators and a large number of their readers, who added their comments and further links to the initial discussion, detailed a growing unease among political webloggers about the typographic features of the memos that CBS had obtained from Burkett.

The memos were written in a proportionally spaced font, not, like most typewriters, in a monospaced font. One memo, dated August 18, 1973, also made use of special characters: a superscript "th" (th) and a single acute accent, frequently used by many people, especially U.S. computer users, as a right quotation mark (′).[66] These characters do not appear on the vast majority of typewriter keyboards from the early 1970s, but could have appeared as part of a customized character set on a typewriter that used a system of swappable typefaces located on daisywheels, like various Royal electric models; or typing balls, such as the IBM Selectric.

For the webloggers, the question of whether the National Guard made use of such machines was only part of the issue. The ensuing debate ranged far and wide, including posts from readers who had been in the military during the same period commenting on the availability of particular machines; links to articles by forensic typewriting experts analyzing the documents; IBM repairmen contesting the claims of other readers about the availability of particular features at the time; links to other weblogs; and so on.[67] Nor was the debate limited to the arcana of typefaces; it expanded into a discussion of the format of military abbreviations, the proper use of memos, and other factors. And, crucially, Powerline was only one node among many in the weblogging network that was hosting this sort of debate. Other weblogs, like the Daily Kos, took a critical and acerbic stance on the Powerline's jingoistic reportage, posting an exhaustive, claim-by-claim refutation on September 10.[68] As with the Powerline debate, the Kos piece raised some provocative questions, but also contained errors of its own. Readers chimed in their own opinions; other webloggers, for and against, blogged away on their own weblogs. As an ever-greater number of weblogs began to cite, link to, and debate each other, they created links that were tracked by search engines like Blogdex, and their sites began to rise in their rankings of current web traffic. As a result, the complexities of the whole debate became more visible to journalists and general readers alike.

While the Powerline administrators declared that the documents were forgeries, it wasn't so much their editorial opinion as the ongoing collective ruckus in the Blogosphere that attracted the attention of larger political websites with paid staff, like Salon.com, then the staff of the major daily newspapers like the *Washington Post*[69] and the *New York Times*,[70] though, as Salon noted, these stories "did not advance the story terribly much beyond what the *Standard* and conservative blogs reported and speculated."[71] The traditional media sources trotted out forensic type experts of their own to deliver their opinions like so many retired Generals during a U.S. invasion, only to be refuted and sometimes dismissed by other sources for their partisan connections. (The two document experts that CBS had hired to do their initial verification "later said they raised red flags that network officials apparently disregarded."[72]) The ruckus continued.

Philosopher Mikhail Bakhtin had a name for this type of discourse, where, as in a conversation or quarrel happening in a group, a number of "utterly incompatible elements" from several different perspectives can dominate by turns without ever resolving into a single perspective: he called it "dialogism."[73] From Bakhtin's perspective, sustained dialogism bootstraps an entire system to "a higher unity" where "the material can develop to the furthest extent what is most original and peculiar in it."[74] What this suggests is that there are significant aspects of Typewritergate that have nothing to do with whether the memos themselves are actually forgeries (and the truth may well never come out). The relationship of typewriting to the production of truth has changed utterly; no one expert opinion sounds convincing one way or the other. Not only are there plenty of other experts with opposing views, but the discursive formation that created the concept of expertise and described the rules by which it functioned has ceded rights to a discursive formation that grants amateurism privilege. Likewise, the monopoly of journalism on the unquestioned production of truth has also weakened, because of the dialogic multiplicity of the Blogosphere. Dan Rather is now perceived as fallible in a way that Walter Cronkite never was, because of the differences in the overall discursive milieu in which they plied the same trade. Rather announced in November

2004 that he would be stepping down as anchor in March of 2005, and subsequently took "personal responsibility" for the story. Though the network took no direct action against Rather, they fired four key employees, including a senior vice-president, the executive producer of *60 Minutes*, and the producer of the actual story.[75] Under these conditions, *even if* we still believed that typewriters still produced truth – and, in case it's not clear, we don't, and they don't – we would be incapable of recognizing it.

Yet Another Ending: The Beginning

This book began with a description of *Royal Road Test*, an art piece that first performs, then begins to map out, the end of typewriting. Pursuing and describing the metaphorical fragments of this abandoned discourse, which only began to become visible after its obsolescence and destruction, leads back to the same point, but with a difference. Sifting through the cultural detritus of typewriting – an archive that includes, but is not limited to, eBay relics, comic books, curatorial pamphlets, web pages, philosophical treatises, photographs, technical manuals, children's books, plays, magazine articles, poems and poetics, zines, trade publications, paintings, memoirs, software, histories, sculptures, economics papers, artists' books, advertisements, novels, instruction manuals, and movies – gradually produces an overall outline of the rules and relations that brought the discourse into being. Piece by piece, the outlines of the machine that produced typewriting become visible, even as the fragments of typewriting itself begin to disappear into the undifferentiated rubble of the past.

Vancouver artist Rodney Graham's installation *Rheinmetall/Victoria 8* (2003)[76] mimes this process off into the infinite. The installation consists of a Victoria 8 movie-projector, made in the 1950s by the Italian firm Cinemeccania, running a continuous loop of film, ten minutes and fifty seconds in length. The film consists of a series of still shots of a Rheinmetall typewriter, manufactured in

Rodney Graham documents typewriting's slow apocalypse.

Germany in the 1930s, gradually being covered in what looks like a fall of snow or very fine ash.

The installation presents a viewer with a series of contrasts: the flatness and silence of the projected images of the typewriter versus the substantial physical presence and noise of the projector. The Victoria 8 is an immense machine; combined with the looping system, it stands about six feet tall. Further, it occupies a simple black platform in the centre of the gallery, which unobtrusively adds another foot to its total height *and* to the perception of it as a "Royal" technology (the name "Victoria" still signifies the zenith of British monarchy, along with a certain gratuitous aesthetic complexity). The noisy clatter of the film running through the projector is the only sound in the room, as the film itself is silent.

The structure of the installation becomes most apparent in the context of the art gallery itself, particularly in the context of a show of Graham's other art.[77] Graham's other video works are usually displayed on TVs connected to DVD players, which pretend to be incidental to the presentation of the work, despite their size. The relative silence of the DVD player makes the equipment necessary to play the videos nearly invisible . . . at least in comparison to the mammoth Victoria 8. As the Wizard of Oz says: Pay no attention to the man behind the curtain. Walking into the room containing *Rheinmetall/Victoria 8* produces an immediate and startling contrast: compared to the relative silence of the rest of the exhibit and the general cathedral-like hushed tones of the art gallery itself, it's *noisy*.

But whatever noise there may be, the first instinct of anyone placed in front of a projector is to look at the projection. The film's subject, the Rheinmetall typewriter, is black, gleaming, pristine, but also obviously anachronistic. Graham found it, of course, in a Vancouver junk shop. He writes, "[It] was in perfect condition. I don't think one word had been typed on it. It was as if it had been preserved in a time capsule."[78] In a manner analogous to the first photo of the Royal Model X in *Royal Road Test*, Graham shoots his film to deliberately emphasize and fetishize the typewriter, rendering it in a scale and mood appropriate for display on a gallery wall: "The shots are individually very long," notes Graham, "long enough for the viewer to imagine he or she is looking at a series of large photographs."[79]

The gradual accumulation of fine white powder on the typewriter (actually flour) signifies some sort of break: a metaphoric winter, as the halcyon days of typewriting slowly recede, or even a nuclear winter after a more apocalyptic break, like the explosion of an atomic weapon (one of the most common visual signifiers of Hiroshima are fragments of machinery from the wreckage, particularly stopped analogue clocks and wristwatches). One way or another, something has changed definitively. As film director Guy Maddin suggests in a discussion of his film *Archangel*, which also takes place during a relentless but gentle fall of obviously ersatz snow, this accumulation of white powder on the hard surfaces of the machine also signals, opiate-like, the passage of an object out of collective memory as it disappears under "a fluffy white blanket of forgetfulness."[80]

The silence of the film evokes the sound-dampening effect that gentle snowfalls have on their immediate environment, but clashes with the racket of the projector (a blanket of "white noise" falling over the viewer as the flour falls over the typewriter). Slowly, by virtue of gradual awareness of the noise that the projector makes, it begins to dawn on the viewer that the projector itself is *also* part of the exhibit, as the installation's title suggests, not just the apparatus of display. Graham writes, "I wanted the two industrial objects to address one another across the space that separates them, two 'obsolete' technologies."[81] Over time, people in the gallery begin to cluster around the machine, to note and appreciate the details of its operation. The "man" behind the metaphorical curtain is not human, but a mechanism with its own particular architecture, rules, and regularities that creates the spectacle on the gallery wall.

This story, of the gradual discovery and explanation of the discursive formations that create and manage the archive of documents and statements on a particular subject at a particular time, is an old story. Plato told it as the myth of the cave centuries before L. Frank Baum wrote *The Wizard of Oz*. It will be told again and again, in different contexts; the film, after all, is a loop, but each reiteration occurs at a different moment in time, and produces different results. If the Victoria 8 component of *Rheinmetall/Victoria 8* functions as an analogy for the discursive formation of typewriting, it also suggests the importance of context. Because of the process of beginning to notice the Victoria 8, the gallery viewer also begins to pay more attention to the DVD players powering the installations in the other galleries, and perhaps, even, to express some curiosity about the gallery itself as a larger machine that envelops and manages the smaller ones it contains, and so on. From a position inside an operating discursive formation, it is difficult, if not impossible, to grasp the total shape of the archives of one's own era, but paying attention to older formations that have gradually become visible may provide some clues.

Here is a final discontinuity: typewriting always implied the presence of at least one person somewhere in the assemblage. Computing does not. When computing disappears, will anyone even notice its absence? And, if so, how will that absence be written? The last typewriter, untouched beneath a fine layer of ash and snow, has nothing to say on the subject. And the loop begins again.

Acknowledgements

This book would never have been completed without the love and support of Frances and Darcy Wershler, Joe Ball, and Caley Strachan.

Thanks to my graduate committee, Marcus Boon, Barbara Crow, Jerry Durlak, Ray Ellenwood, and Will Straw, for shepherding the first incarnation of this project into existence. Special thanks to my supervisor, Terry Goldie, for his patience and kindness.

Hadley Dyer and Hal Niedzviecki went to the trouble of reading the entire monstrous manuscript, and provided valuable professional insight during the editing phase.

Thanks to Hilary McMahon, Nicole Winstanley, and the rest of Westwood Creative Artists for placing this book in capable editorial hands. My editor, Alex Schultz, patiently helped me pare three hundred pages from the manuscript; thanks to him, and everyone at M&S, for believing in this book.

Thanks to all those who provided suggestions and leads during the research phase of this book, especially Ken Babstock, Stan Bevington, Jack David, Johanna Drucker, Nicky Drumbolis, Paul Dutton, Craig Dworkin, Sandra Gabriele, Michael Helm, Neil Hennessy, Karen Mac Cormack, Steve McCaffery, Ellie Nichol, Kent Nussey, Marjorie Perloff, Alex Pugsley, Rick/Simon, Marvin Sackner, Lisa Sloniowski, and everyone who ever told me a story about the old typewriter in their basement.

Special thanks to my friends and co-conspirators, Bruce Andrews, Charles Bernstein, Christian Bök, Rob Fitterman, Kenneth Goldsmith, Bill Kennedy, Michael Redhill, Brian Kim Stefans, and Alana Wilcox.

To the poets: there are other books to be written about typewriting. At least one of them will be about typewritten concrete and visual poetry, because I'll be writing it next . . .

Permissions

(Every effort has been made to secure all permissions required for reproduction of images in this book. In the case of any error or omission, a correction will be made in a subsequent printing upon notification.)

Page 13, Royal Road Test image courtesy of Ed Ruscha; **page 30**, Sam Messer painting courtesy of Distributed Art Publishers, Inc; **page 58**, Francis machine image courtesy of Dover Publications, Inc; **page 64**, Pratt pterotype image courtesy of Dover Publications, Inc; **page 111**, Burroughs image courtesy of Sterling Lord Literistic, Inc; **page 116**, Burroughs image courtesy of City Lights Books; **page 122**, Bugwriter images courtesy of David Cronenberg Productions; **page 126**, Mujahideen typewriter image courtesy of David Cronenberg Productions; **page 182**, Key images from Hiss machine courtesy of Harper & Row (HarperCollins); **page 201**, Betsy Lewin illustration courtesy of Simon & Schuster, Inc; **page 207**, Borgese images courtesy of Holt, Rinehart and Winston; **page 213**, George Herriman cartoon courtesy of Random House; **page 215**, George Herriman cartoon courtesy of Random House; **page 223**, Northrop Frye image reproduced from John Ayre, *Northrop Frye: A Biography* (Toronto: Random House of Canada, 1989), original source unknown; **page 272**, Hebern rotor image courtesy of Pan Macmillan; **page 277**, Geof Darrow illustration courtesy of DC Comics; **page 287**, Rodney Graham image courtesy of Donald Young Gallery.

Notes

Introduction

1. Blackwood, Algernon. "A Psychical Invasion." *John Silence: Physician Extraordinary*. London: Eveleigh Nash/Fawside House, 1908. 1–71.
2. Ibid., 7.
3. Ibid., 21.
4. Ibid., 16.
5. Ibid., 24.
6. Ibid., 21, 25.
7. Ibid., 29.
8. Ibid., 27.
9. Ibid., 28.
10. Ibid., 3.
11. Ibid., 36.
12. Ibid., 69.
13. Ibid.
14. Ibid., 15.
15. Ibid., 70.
16. Ibid., 69–70.
17. *Now* 1199 v. 24. 19 (January 6–12, 2005): 68.
18. Virilio, Paul. *Unknown Quantity*. New York: Thames & Hudson/Fondation Cartier pour l'art contemporain, 2003. 5.

Part 1

Chapter 1

1. Ruscha, Edward, Mason Williams, Patrick Blackwell. *Royal Road Test.* [1967]. Los Angeles, 1980. 4th edn. No pagination. All subsequent references to *Royal Road Test* are from this edition.
2. A powerful enough image that it later became the logo for Tarantino's production company, A Band Apart <www.abandapart.com>.
3. Available in many places online and in print, including <parazite.host.sk/cia.html>.
4. Stallabrass, Julian. *Gargantua: Manufactured Mass Culture.* London: Verso, 1996. 176.
5. Ibid., 177–78.
6. Ibid., 178.

Chapter 2

1. Foucault, Michel. *The Archaeology of Knowledge and the Discourse on Language.* [1969]. Trans. A. M. Sheridan Smith. New York: Pantheon Books, 1972. 130.
2. <www.williamgibsonbooks.com/archive/2003_03_15_archive.asp>
3. <www.wired.com/wired/archive/7.01/ebay.html >
4. Niedzviecki, Hal. *We Want Some Too: Underground Desire and the Reinvention of Mass Culture.* Toronto: Penguin, 2000. 66–69.
5. Ibid., 293.
6. Ibid., 160.
7. <www.ebay.com>. As eBay pages are generated dynamically and erased after their time-based auctions conclude, there is no record of these items, nor would a URL be of any real use.

Chapter 3

1. Shenk, Joshua Wolf. "The Things We Carry." *Harper's Magazine*, June 2001: 46–56.
2. Ibid., 46.
3. Baudrillard, *System*, 75.
4. Shenk, 46.
5. Baudrillard, *System*, 74.
6. Ibid., 74.
7. Shenk, 50–51.

8. Shenk, 55.
9. Baudrillard, *System*, 55.
10. Ibid., 76.
11. <www.guinessworldrecords.com/content_pages/record.asp?recordid=54002>
12. Ibid., 56.
13. Jameson, Fredric. *Postmodernism, or, The Cultural Logic of Late Capitalism*. Durham: Duke University Press, 1991. 19.
14. Ibid., 156.
15. Sedaris, David. "Nutcracker.com." *Me Talk Pretty One Day*. Boston/New York/London: Little, Brown and Company, 2000. 142–49.
16. Ibid., 146.
17. Ibid., 147.
18. Baudrillard, Jean. *Revenge of the Crystal: Selected Writings on the Modern Object and Its Destiny, 1968–1983*. Ed. and Trans. Paul Foss and Julian Pefanis. London; Concord: Pluto Press, 1990. 35, 36.
19. Ibid., 37.
20. Knight, Stephen. "The Weight of Words." *Globe and Mail*, Tues 5 Aug 2003. A16.
21. Baudrillard, *System*, 77.
22. Ibid., 79.
23. Auster, Paul. Illus. Sam Messer. *The Story of My Typewriter*. New York: D.A.P., 2002.
24. Ibid., 9.
25. Ibid., 15.
26. Ibid., 21.
27. Ibid., 22.
28. Ibid., 22.
29. Ibid., 16.
30. Ibid., 17.
31. Ibid., 22–23.
32. Ibid., 23.
33. Ibid., 28–29.
34. Ibid., 32.
35. Ibid., 28.
36. Ibid., 55–56.

Part 2

Chapter 4

1. Beeching, Wilfred A. *Century of the Typewriter*. London: Heinemann, 1974. 1.
2. ———. *Foucault Live (Interviews, 1966–84)*. Trans. John Johnston. Ed. Sylvère Lotringer. New York: Semiotext(e), 1989. 47. See also Foucault, Michel. "On the Archaeology of the Sciences." *Aesthetics, Method, and Epistemology: Essential Works of Foucault 1954–1984*. [1994]. Vol. 2. Ed. James D. Faubion. Trans. Robert Hurley et al. New York: The New Press, 1998. 297–333. 306.
3. Beeching, 28; Adler, 136.
4. Ibid., 3.
5. <en.wikipedia.org/wiki/Pantograph>
6. Beeching, 7.
7. <www.monticello.org/reports/interests/polygraph.html>
8. Ibid.
9. McLuhan, Marshall. *Understanding Media: The Extensions of Man*. New York: Signet Books, 1964.
10. "Whim." *The Compact Edition of the Oxford English Dictionary*. 25th printing. 1971.
11. Richards, G. Tilghman. *The History and Development of Typewriters*. London: Her Majesty's Stationery Office, 1964. 4–7.
12. Richards, 9.
13. Beeching, 8–9; Adler, 60–61.
14. Richards, 18.
15. Beeching, 10; Adler, 62–63.
16. Adler, 78.
17. Adler, 78–79; "The First American Type Writer." *Scientific American*. Vol. LVL. no. 18. 30 April 1887.
18. Beeching, 11.
19. Beeching, 12.
20. Russo, Thomas A. *Mechanical Typewriters: Their History, Value and Legacy*. Atglen: Schiffer Publishing Co., 2002. 9; Adler, 93.
21. Russo, 12; Adler, 99.
22. Russo, 10.
23. I am indebted to Steve McCaffery for suggesting the Christopher Smart example.

24. Smart, Christopher. *Jubilate Agno*. Fragment B, Part 3.
25. Kittler, 241.
26. "All typewriter history commences with a British patent granted to a Henry Mill . . . ," Richards, 17.
27. Adler, 48.
28. Beeching, 5; Adler, 60.
29. Ibid., 7.

Chapter 5

1. Beeching, 7.
2. Richards, 17.
3. Adler, 168.
4. Beeching, 8.
5. <www.kevinlaurence.net/essays/cc.shtml>. From Adler, Michael H. *The Writing Machine*. London: Allen & Unwin, 1973.
6. Beeching, 8.
7. Kittler, 188.
8. Richards, 21.
9. Beeching, 13.
10. Adler, 68–70.
11. Richards, 20.
12. Ibid., 12.
13. Qtd. in Seltzer, Mark. *Bodies and Machines*. New York: Routledge, 1992. 11.
14. Kittler, Friedrich. *Gramophone, Film, Typewriter*. Trans. Geoffrey Winthrop-Young and Michael Wutz. Stanford: Stanford University Press, 1999. 189.
15. Kittler. Beeching uses "Rasmus Hans"; I generally have opted for Kittler's spellings because he sources his information from a Danish source: Nyrop, Camillus. 1938. "Malling Hansen." In *Dansk Biografisk Leksikon*, ed. Povl Engelstoft, vol. 18. Copenhagen. 265–27.
16. Beeching, 22.
17. Richards, 7.
18. Burghagen, qtd. in Kittler 202.
19. Richards, 7.
20. Qtd. in Kittler 202. Burghagen, Otto. *Die Schreibmaschine. Illustrierte beschreibung aller gangbaren Schreibmaschinen nebst gründlicher Anleitung zum Arbeiten auf sämlichen Systemen*. Hamburg. 6.

21. Beeching, 22–23; Adler, 157–57.
22. Nietzsche, Friedrich. *Daybreak: Thoughts on the Prejudices of Morality.* [1881]. Ed. Maude Marie Clark and Brian Lieter. Trans. R.J. Hollingdale. Cambridge: Cambridge University Press, 1997. 30.
23. Kittler, 200–01.
24. Adler, 154.
25. Qtd. in Kittler, 200.
26. Ibid., 202.
27. Ibid., 207.
28. Dvorak, August. *Typewriting Behavior: Psychology Applied to Teaching and Learning Typewriting*. New York: American Book Company, 1936. 74.

Chapter 6

1. <www.nyu.edu/pages/linguistics/courses/v610051/gelmanr/cult_hist/text/p298.html>; Adler, 50.
2. <www.miralab.unige.ch/subpages/automates/eightennth/knaus_uk.htm>; Adler, 50.
3. Adler, 51.
4. <en.wikipedia.org/wiki/The_Turk>
5. Standage, Tom. *The Turk: The Life and Times of the Famous Eighteenth-Century Chess-Playing Machine*. New York: Walker & Company, 2002.
6. <www.wired.com/wired/archive/10.03/turk_pr.html>
7. <www.haskins.yale.edu/haskins/HEADS/SIMULACRA/kempelen.html>
8. <www.slovakradio.sk/radioinet/kultura/expstudio/kempe.html>
9. Beeching, 7.
10. Bliven, 46; Adler, 48.
11. Harrison, John. *A Manual of the Type-Writer*. London: Isaac Pitman, 1888. 9.
12. Qtd. in Beeching, 3.
13. Ibid., 9; Adler, 67.
14. Richards, 21–22.
15. Adler, 99–100.
16. Richards, 18–19.
17. Ibid., 19.
18. Beeching, 14.
19. Richards, 19; Adler, 72.

20. Beeching, 14; Adler, 129.
21. Richards, 19.
22. Ibid., 26.
23. Beeching 23–24; Adler, 106–10.
24. Foucault, Michel. *The Archaeology of Knowledge and the Discourse on Language.* [1969]. Tr. A. M. Sheridan Smith. New York: Pantheon Books, 1972. 38.

Chapter 7

1. Beeching, 21; Adler, 130–35.
2. Beeching, 21.
3. Beeching, 21; Adler, 131.
4. Kittler, 190, and 291 n.26, which refers to Granichstaedten-Czerva, Rudloph von. *Peter Mitterhofer, Erfinder der Schreibmaschine: Ein Liebsbild.* Wien, 1924. 35.
5. Kittler, 190.
6. Ibid., 20.
7. Ibid., 24–25.
8. <www2.milwaukee.k12.wi.us/sholes/>
9. Beeching, 28; Adler, 137
10. Richards, 24–25.
11. Qtd. in Richards, 23; Adler, 138.
12. Bliven, 44.
13. Ibid.
14. Richards, 23; Adler, 138.
15. Bliven, Bruce, Jr. *The Wonderful Writing Machine.* New York: Random House, 1954. 46.
16. Beeching, 32; Adler, 125.
17. Herkimer 34.
18. Beeching, 29.
19. Richards, 23; Adler, 160.
20. Beeching, 28–29; Adler, 368.
21. Richards, 23; Adler, 143.
22. Beeching, 29.
23. Ibid., 30.
24. Richards, 23.
25. Beeching, 29.
26. Ibid., 32.

27. Adler, 146.
28. Richards, 23–24; Adler, 146–47.
29. Beeching, 31; Adler, 148.
30. Beeching, 32.
31. Bliven, 56.
32. Richards, 24.
33. Ibid., 24.
34. Beeching, 32.
35. Ibid., 27.
36. Richards, 24.
37. Russo, 19.

Part 3

Chapter 8

1. Ben'Ary, Ruth. *Touch Typing in Ten Lessons: A Home-Study Course with Complete Instructions in the Fundamentals of Touch Typewriting*. [1945]. Revised Edition. New York: Perigee Books, 1989. 10.
2. *The Compact Edition of the Oxford English Dictionary*. Oxford: Oxford University Press, 1971. 25th U.S. printing. 66.
3. Benveniste, Emile. *Problems in General Linguistics*. [1966]. Trans. May Elizabeth Meek. Coral Gables: University of Miami Press, 1971. 227.
4. Ronell, Avital. *Dictations: On Haunted Writing*. Lincoln: University of Nebraska Press, 1989. 89.
5. Ibid., 78.
6. Kittler, 223.
7. Derrida, Jacques. *The Post Card: From Socrates to Freud and Beyond*. [1980]. Trans. Alan Bass. Chicago: The University of Chicago Press, 1987.

Chapter 9

1. McLuhan, Marshall. *Understanding Media: The Extensions of Man*. New York: Signet Books, 1964. 230.
2. Bliven, Bruce, Jr. *The Wonderful Writing Machine*. New York: Random House, 1954. 72.
3. Thayer, William Roscoe. *Theodore Roosevelt: An Intimate Biography*. Boston: Houghton Mifflin, 1919. New York: Bartleby.com, 2000. <www.bartleby.com/170/14.html>

4. "I know one *president* whose staff consists of two typists." Mencken, H. L. *The American Language*. 2nd edn. New York: Alfred A. Knopf, 1921. New York: Bartleby.com, 2000. <www.bartleby.com/185/21.html>

Chapter 10

1. Wicke, Jennifer. "Vampiric Typewriting: *Dracula* and Its Media." *ELH* 59.2 (summer 1992): 467–93.
2. Ibid., 477.
3. Kittler, 221.
4. Stoker, Bram. *Dracula*. [1897]. New York; Signet Classic, 1965. 380.
5. Ibid., 286.
6. Ibid., 224.
7. Wicke, 476.
8. Stoker, 55.
9. Ibid., 223.
10. Wicke, 471.
11. Ibid., 474.
12. Stoker, 288.
13. Ibid., 289.
14. Wicke, 485.
15. Ibid., 490.
16. Ibid., 492.

Chapter 11

1. Kittler, 184.
2. From the U.S. Bureau of the Census, *Sixteenth Census of the United States, 1940: Population* (1943). As cited in Davies, Margery. *Woman's Place Is at the Typewriter: The Feminization of the Clerical Labour Force*. Somerville, Mass, 1974. 10. In Kittler, 184.
3. Beeching, 35.
4. Ibid., 35.
5. Keep, Christopher. "The Cultural Work of the Type-Writer Girl." *Victorian Studies* 40:3. <iupjournals.org/victorian/vic40-3.html>
6. McLuhan, 228.
7. Ibid., 187–88.
8. Herkimer, 9.
9. Ibid., 140.

10. Current, Richard N. *The Typewriter and the Men Who Made It.* Urbana, 1954. 86.
11. In Keep.
12. Ibid.
13. Beeching, 35.
14. Kipling, Rudyard. *From Sea to Sea: Letters of Travel, Vol. II.* Vol. XVI of *The Works of Rudyard Kipling.* 25 vols. New York: Charles Scribner, 1913. 80. Qtd. in Keep.
15. Allen, Grant ["Olive Pratt Rayner"]. The Type-Writer Girl. London: C. Arthur Pearson, 1897. Qtd. in Keep.
16. Barrie, J. M. "The Twelve-Pound Look." *The Twelve-Pound Look and Other Plays.* London: Houghton, 1928. 3–43. 29. Qtd. in Keep.
17. Ibid., 40. Qtd. in Keep.
18. Seaton, George, and Edmund Goulding (uncredited). *The Shocking Miss Pilgrim.* 1947. Sadly, this film has never been released on VHS or DVD.
19. <imdb.com/title/tt0039819/?fr=c2l0ZT1kZnxodDoxfGZiPXV8cG49 MHxxPXRoZSBzaG9ja2luZyBtaXMgcGlsZ3JpbXxteDoyMHxsbTo1 MDB8aHRtbDox;fc=1;ft=3;fm=1#comment>
20. Reproduced in Russo, 246.
21. Maas, Frederica Sagor. *The Shocking Miss Pilgrim: A Writer in Early Hollywood.* Lexington: University of Kentucky Press, 1999.
22. <www.salon.com/people/feature/1999/08/13/maas/print.html>
23. Hutchins, B. L. "An Enquiry Into the Salaries and Hours of Work of Typists and Shorthand Writers." *The Economic Journal* 16 (1906): 445–49. Qtd. in Keep.
24. Courtney, Janet Elizabeth Hogarth. *Recollected in Tranquillity.* London: Heinemann, 1926. 147–48. Qtd. in Keep.
25. In Keep.
26. Stoker, 236.
27. Price, 218.
28. Ibid.
29. Beeching, 34.
30. Bliven, 72.
31. <tijuanabibles.org/cgi-bin/hazel.cgi?action=detail&item=TB057>; <tijuanabibles.org/cgi-bin/hazel.cgi?action=detail&item=TB037>
32. In Keep.

Chapter 12

1. Beeching, 36–37.
2. *The Typewriter: An Illustrated History*. Mineola: Dover Publications, 2000. Reprint of *The Typewriter: History and Encyclopedia*. New York: Typewriter Topics, 1924. 16.
3. Price, Leah. "From Ghostwriter to Typewriter: Delegating Authority at Fin de Siècle." *The Faces of Anonymity: Anonymous and Pseudonymous Publications for the Sixteenth to the Twentieth Century*. Ed. Robert J. Griffin. New York: Palgrave Macmillan, 2003. 211–32. 214.
4. Ibid., 215.
5. Ibid., 217.
6. Qtd. in Seltzer, Mark. *Bodies and Machines*. New York: Routledge, 1992. 195 n. 57.
7. Bosanquet, Theodora. *Henry James at Work*. London: Hogarth Press, 1924.
8. Ibid., 243.
9. Ibid.
10. Bosanquet, 244; James, Henry. *The Letters of Henry Jones*. Ed. Perry Lubbock. vol II. London: MacMillan and Co., Ltd. 1920. 212.
11. McLuhan, *Understanding Media*, 229.
12. Ibid.
13. Bosanquet, 245.
14. Ibid., 247.
15. Ibid., 248.
16. Ibid.
17. McLuhan, *Understanding Media*, 229.
18. Bosanquet, 248.
19. Ibid.
20. Kittler, 214.
21. Ibid.
22. Ibid., 216.
23. Ibid.
24. Thurschwell, Pamela. "Henry James and Theodora Bosanquet: On the Typewriter, in *The Cage*, at the Ouija Board." *Textual Practice* 13(1), 1999. 5–23. 11, 15.
25. Ibid., 11.
26. Ibid., 12.

27. Ibid., 13.
28. Ibid., 14.
29. Ibid., 14.
30. Ibid., 17.

Chapter 13

1. Morgan, Ted. *Literary Outlaw: The Life and Times of William S. Burroughs*. New York: Henry Holt and Company, 1988. 15.
2. Ibid., 16.
3. Ibid., 17–18.
4. Russo, Thomas A. *Mechanical Typewriters: Their History, Value, and Legacy*. Atglen: Schiffer Publishing, 2002. 96.
5. Corradini, Bruno, and Emilio Settimelli. "Weights, Measures and Prices of Artistic Genius." 1914. *Futurist Manifestoes*. Ed. Umbro Appolonio. Trans. Robert Brain, R. W. Flint, J. C. Higgitt, Caroline Tisdall. The Documents of 20th Century Art. New York: The Viking Press, 1973. 135–50. 136.
6. Lemaire, Gérard-Georges. "23 Stitches Taken by Gérard-Georges Lemaire and 2 Points of Order by Brion Gysin." In Burroughs, William S., and Brion Gysin. *The Third Mind*. New York: Seaver Books, 1979. 9–24. 16.
7. Burroughs, William S. "Inside the Control Machine." Burroughs, William S., and Brion Gysin, *The Third Mind*. 178–79.
8. Burroughs, William S. *Nova Express*. 1964. Reprint edn. New York: Grove Press, 1992. 66, 85.
9. ———. *The Ticket That Exploded*. 1962. Reprint edn. New York: Grove Press, 1992. 159.
10. A contemporary of Burroughs, Broyard was a New York bookstore owner, essayist, and literary critic for the *New York Times* who took a dim view of Burroughs's writing. As a result, an abject parody of Broyard makes frequent appearances throughout Burroughs's work. As a light-skinned African-American who frequently "passed" for white while displaying an open dislike of people with darker skin, Broyard was in fact struggling with contradictory narratives overwritten on his skin by the Soft Typewriter.
11. ———. *My Education: A Book of Dreams*. New York: Viking, 1995. 160.
12. Ibid., 113.
13. Burroughs, William S. *The Ticket That Exploded*, 160.

Chapter 14

1. Petroski, Henry. *The Pencil: A History of Design and Circumstance*. New York: Alfred A. Knopf, 1990. 4.
2. Burroughs, William S. "Technology of Writing." *The Adding Machine: Collected Essays*. London: John Calder, 1985. 37.
3. Tzara, Tristan. "Dada Manifesto on Feeble Love and Bitter Love." *Seven Dada Manifestoes and Lampisteries* [1963]. Trans. Barbara Wright. London: Calder Publications, 1977. 39.
4. Ibid., 184.
5. Burroughs, William S., and Brion Gysin. *The Third Mind*. 29.
6. Gysin, Brion, and Terry Ward. *Here to Go*, 51.
7. Burroughs, William S. *The Job*, 30.
8. Burroughs, William S., and Brion Gysin. *The Third Mind*. 32.
9. Ibid., 193.
10. Ibid., 54.
11. Ibid., 98.
12. Gysin, Brion, to Rober Palmer. *Rolling Stone* May 1972. Qtd. in Gysin, Brion, and Terry Ward. *Here to Go*, 55.
13. Burroughs, William S. *The Job*, 28.
14. Burroughs, William S. *Naked Lunch*. 1959. Reprint edn. New York: Grove Press, 1992. 86.
15. Burroughs, William S. *The Place of Dead Roads*. 1983 Reprint edn. New York: Picador, 2001. 225.
16. Ibid., 96.
17. Burroughs, William S. *Queer*. 1951. New York: Penguin Books, 1995. Introduction.
18. Burroughs, William S. "It Belongs to the Cucumbers." *The Adding Machine: Collected Essays*. London: John Calder, 1985. 53–60. 55.
19. Ibid., 60.
20. Ibid., 55.
21. Lemaire, Gérard-Georges, 187.
22. Gysin, Brion, and Terry Ward. *Here to Go*, 187–88.
23. Ibid., 188.
24. Ibid., 189.
25. Burroughs, William S. *The Job*, 164.
26. *Nova Express*, 152.
27. Ibid., 118.
28. Ibid., 116.

Chapter 15

1. Cronenberg, David. *Naked Lunch*. Toronto: Alliance Films, 1991. Criterion Collection no. 220.
2. Cronenberg, David. "Back to the Future: Making *Naked Lunch*." Ed. Chris Rodley. *Cronenberg on Cronenberg*. Toronto: Alfred A. Knopf, 1992. 157–70. 162.
3. Cronenberg, David. *Naked Lunch*. Director's Commentary, ch. 8.
4. Ibid., Ch. 9.
5. Ronell, Avital. *Crack Wars: Literature, Addiction, Mania*. Lincoln: University of Nebraska Press, 1992. 33.
6. Ibid., 51.
7. Boon, Marcus. *The Road of Excess: A History of Writers on Drugs*, Cambridge: Harvard University Press, 2002.
8. Burroughs, William S. *Naked Lunch*. Introduction, xxxv.
9. Burroughs, William S. *The Job*, 159.
10. Burroughs, William S., and Brion Gysin. *The Third Mind*, 43.
11. Ibid., 159.
12. Boon, Marcus, 164–65.
13. Burroughs, William S. "The Name Is Burroughs." *The Adding Machine*, 1–18. 11.
14. Cronenberg, David. *Naked Lunch*. Director's Commentary, Ch. 5.
15. Cronenberg, David. *Naked Lunch*. Ch. 9.
16. Cronenberg, David. "Back to the Future: Making *Naked Lunch*." 165.
17. Cronenberg, David. *Naked Lunch*. Director's Commentary, Ch. 9.
18. Cronenberg, David. *Naked Lunch*. Ch. 9.
19. Ibid., Ch. 11.
20. Ibid., Ch. 12.
21. See esp. Burroughs, William S., *The Job*, 165–66.
22. Cronenberg, David. *Naked Lunch*. Ch. 8.
23. Ibid., Ch. 10.
24. Ibid., Ch. 13.
25. Cronenberg, David. *Naked Lunch*. Director's Commentary, Ch. 19.
26. Cronenberg, David. *Naked Lunch*. Ch. 13.
27. Ibid., Ch. 14.
28. Ibid., Ch. 11.
29. Ibid., Ch. 14.
30. Ibid., Ch. 13.
31. Said, Edward. *Orientalism*. New York: Vintage Books, 1978. 12.

32. Ibid., 17.
33. Herkimer County Historical Society. *The Story of the Typewriter 1873–1923*. Herkimer, New York: Herkimer County Historical Society, 1923. 128.
34. Ibid.
35. Cronenberg, David. *Naked Lunch*. Director's Commentary, Ch. 13.
36. Russo, Thomas A. *Mechanical Typewriters: Their History, Value and Legacy*. Atglen: Shiffer Publishing, 2002. 170.
37. <en.wikipedia.org/wiki/Mujahideen>
38. This is the substance that Tom Frost misrepresented to Lee as majoun. A hashish paste, majoun is ingested; Lee gingerly applies this substance to a spot over one of the arteries on his neck, where it has created a bruise or stain.
39. Oxenhandler, Neil. "Listening to Burroughs's Voice." *William S. Burroughs at the Front: Critical Reception, 1959–1989*. Ed. Jennie Skerl and Robin Lyndenberg. Carbondale: South Illinois University Press, 1991. 133–47. 142.
40. Cronenberg, David. *Naked Lunch*. Director's Commentary, Ch. 5.
41. Ibid., Ch. 7.
42. Cronenberg, David. *Naked Lunch*. Ch. 20.
43. Ibid., Ch. 19.
44. Ibid., Ch. 20.

Part 4

Chapter 16

1. Herkimer County Historical Society. *The Story of the Typewriter 1873–1923*. Herkimer, New York: Herkimer County Historical Society, 1923. 9.
2. Ibid., 14.
3. Qtd. in Beeching, Wilfred A. *Century of the Typewriter*. London: Heinemann, 1974. 24.
4. Herkimer County Historical Society. *The Story of the Typewriter*, 48–49.
5. <www.blonnet.com/businessline/2000/10/18/stories/041855ju.htm>
6. Heidegger, Martin. *Parmenides*. Trans. André Schuwer and Richard Rojcewicz. Bloomington: Indiana University Press, 1992. 80–81.
7. Beeching, 124.

8. Bliven, 202.
9. Ibid., 204.
10. Seltzer, Mark. *Bodies and Machines*. New York: Routledge, 1992. 157, 159.
11. Ibid.
12. Crowe, Cameron. *Say Anything*. Culver City: Gracie Films, 1989.
13. Foucault, Michel. "Truth and Power." *Power: Essential Works of Foucault 1954–1984*. Vol. 3. Ed. Paul Rabinow. Trans. Robert Hurley and others. New York: The New Press, 1994. 111–33. 125.
14. Foucault, Michel. *Discipline and Punish: The Birth of the Prison*. [1975]. Trans. Alan Sheridan. New York: Vintage Books, 1979. 137.
15. de la Mettrie, Julien Offray. *Man a Machine*. [1748]. Trans. Gertrude C. Bussey and others. La Salle, Open Court Classics, 1987.
16. Foucault, Michel. *Discipline and Punish: The Birth of the Prison*. 136.
17. Ibid., 136–37.
18. Ibid., 141.
19. Ibid., 143.
20. Ibid., 145–46.
21. Ibid., 149–50.
22. Ibid., 154.
23. Ibid., 151.
24. Ibid., 152.
25. Ibid., 153.
26. Ibid., 161.
27. Mogyorody. *Typing 100*. Toronto: McGraw Hill, 1970.
28. Ibid., 4.
29. Ibid., 123.
30. Ford, Henry. *My Life and Work*. Garden City: Doubleday, 1923. 108–9.

Chapter 17

1. <gilbrethnetwork.tripod.com/bio.html>
2. <www.armchair.com/warp/hf3.html>
3. Gilbreth, Frank B., and Ernestine Gilbreth Carey. *Cheaper by the Dozen*. [1948]. New York: Perennial Classics, 2002.
4. Ibid., 1.
5. Ibid., 2.
6. Powell, Michael. *Peeping Tom*. London, 1960. Criterion Collection 58.
7. Gilbreth, Frank B., and Ernestine Gilbreth Carey. *Cheaper by the Dozen*. 2.

8. Gilbreth, Frank B., and Ernestine Gilbreth Carey. *Cheaper by the Dozen*. 42.
9. Gilbreth, Frank B., and Ernestine Gilbreth Carey. *Cheaper by the Dozen*. 43.
10. Ibid., 44.
11. Ibid., 42.
12. Ibid.
13. Ibid.
14. Ibid., 44.
15. Ibid.
17. Ibid., 45.
17. Ibid., 140.

Chapter 18

1. Gilbreth, Frank B., and Ernestine Gilbreth Carey, 2.
2. Herkimer County Historical Society. *The Story of the Typewriter*, 68.
3. "Factoid." *The Canadian Oxford Dictionary*. Toronto: Oxford University Press, 1998.
4. Herkimer County Historical Society. *The Story of the Typewriter*, 11.
5. Beeching, 40.
6. Russo, 15, 19.
7. Bliven, Bruce, Jr. The Wonderful Writing Machine. New York: Random House, 1954. 143.
8. Beeching, 38.
9. Lundmark, Torbjörn. *Quirky QWERTY: A Biography of the Typewriter and its Many Characters*. New York: Penguin Books, 2002. 19.
10. See <members.aol.com/alembicprs/selcase.htm#usc> for a set of case lay schemes organized by name and date.
11. See for example Richards 24, Beeching 39.
12. Richards, 24.
13. Bliven, 145.
14. Ibid.

Chapter 19

1. Cooper, 6.
2. Beeching, 41.
3. Dvorak, 209.

4. Liebowitz, S. J., and Stephen E. Margolis. "The Fable of the Keys." Originally published in *Journal of Law & Economics* vol. XXXIII (April 1990). <www.utdallas.edu/~liebowit/keys1.html>
5. Liebowitz and Margolis provide the following citations: David, Paul A. "Clio and the Economics of QWERTY." *Am Econ. Rev.* 75 (1985). 332; and "Understanding the Economics of QWERTY: The Necessity of History." *Economic History and the Modern Economist*. Ed. William N. Parker, 1986.
6. Ibid.
7. Ibid.
8. Ibid.
9. Ibid.
10. Ibid.
11. Ibid.
12. Ibid.
13. Ibid.
14. Cooper, 7.

Chapter 20

1. Bliven, 43.
2. Ibid., 68.
3. McLuhan, Marshall. *Understanding Media: The Extensions of Man*. New York: Signet Books, 1964. 227–28.
4. Olson, Charles. "Projective Verse." *20th Century Poetry and Poetics*. Ed. Gary Geddes. 2nd ed. Toronto: Oxford University Press, 1973. 526–38. 533.
5. Olson, Charles, 534.
6. Ibid., 536.
7. Ibid., 536.
8. Ibid., 527.
9. McLuhan, Marshall. *Counterblast*. Toronto: McClelland and Stewart, 1969. 16–17.
10. Ibid.
11. Ibid., 535.
12. Ibid., 532.
13. Ibid., 534.
14. <wings.buffalo.edu/epc/ezines/passages/passages5/forster.htm>
15. Ibid.

16. Ibid.
17. Eigner, Larry. “Q & As (?) Large and Small: Parts of a Collaborate.” *areas lights heights: writings 1954–1989*. Ed. Benjamin Friedlander. New York: Roof Books, 1989. 148–66. 150.
18. Ibid., 149.
19. Ibid.
20. Ibid.
21. Sudnow, David. *Talk’s Body*. [1979]. Middlesex: Penguin Books, 1980. Preface.
22. Ibid., 25–26.
23. Ibid., 11.
24. Foucault, Michel. *Discipline and Punish*, 172–77.
25. Sudnow, 10.
26. Ibid., 11.
27. Ibid., 49.
28. Derrida, Jacques. “Signature Event Context.” *Margins of Philosophy*. [1972]. Trans. Alan Bass. Chicago: University of Chicago Press, 1982. 329–30.
29. Sudnow, 17.
30. Ibid., 98.
31. Ibid., 124.

Chapter 21

1. Derrida, Jacques. *Of Grammatology*. [1967]. Trans. Gayatri Chakravorty Spivak. Baltimore: The Johns Hopkins University Press, 1976. 7, 11, 12, 17 et passim.
2. Doyle, Arthur Conan. “A Case of Identity.” [1891]. *The Illustrated Sherlock Holmes Treasury*. Revised and expanded. New York: Avanel Books, 1984. 39.
3. Foucault, Michel. “Truth and Power.” *Power: Essential Works of Foucault 1954–1984*. Vol. 3. Ed. Paul Rabinow. 111–33. 131.
4. <homepages.nyu.edu/~th15/bradford.html>
5. <faculty.ncwc.edu/toconnor/315/315lect05.htm>
6. Ibid.
7. Ibid.
8. Ibid.
9. Ibid.
10. <homepages.nyu.edu/~th15/who.html>

11. Ibid.
12. Ibid.
13. Ibid.
14. Ibid.
15. Ibid; ; see also Hiss, Alger. *In the Court of Public Opinion*. New York: Harper & Row, 1957.
16. Qtd. in Hiss, 403.
17. Hiss, 403–04.
18. Qtd. in Hiss, 405.
19. <homepages.nyu.edu/~th15/typewr.html>
20. <homepages.nyu.edu/~th15/greenreuben.html>

Chapter 22

1. <en.wikipedia.org/wiki/Infinite_monkey_theorem>
2. Maloney, Russell. "Inflexible Logic." [1940]. In Newman, James R., ed. *The World of Mathematics*. Vol. 4. New York: Simon & Schuster, 1956. 2262–67. <kimura.tau.ac.il/graur/Texts/logic.htm>
3. <www.remingtonsociety.com/questions/Smoot.htm>
4. Russo, 21.
5. <www.tvtome.com/tvtome/servlet/GuidePageServlet/showid-146/epid-1361>
6. Russo, 21.
7. Adams, Douglas. *The Hitch Hiker's Guide to the Galaxy: A Trilogy In Four Parts*. [1979]. London: Heinemann, 1984. 68.
8. <en.wikipedia.org/wiki/Infinite_monkey_theorem>
9. Hendrickson, Robert. *The Literary Life and Other Curiosities*. New York: Penguin Books, 1981. 160.
10. ReMine, W. J. *The Biotic Message: Evolution versus Message Theory*. Saint Paul: St. Paul Science, 1993. Qtd. in Hennessy, Neil. "Willy vs. Jimmy: Which Typing Monkey gets all the Bananas?" [2003]. PDF privately published 28 June, 2004, "in honour of Dr. Christiam oBok's [sic] departure from Toronto to assume his position as Projector of the Academy."
11. <user.tninet.se/~ecf599g/aardasnails/java/Monkey/webpages/index.html#results>
12. <user.tninet.se/~ecf599g/aardasnails/java/Monkey/webpages/#results>
13. <education.guardian.co.uk/higher/sciences/story/0,12243,952359,00.html>

14. <www.vivaria.net/experiments/notes/documentation/>
15. <education.guardian.co.uk/higher/sciences/story/0,12243,952359,00.html>
16. <www.vivaria.net/experiments/notes/publication/>
17. Ibid.
18. Ibid.
19. Ives, David. *Words, Words, Words*. [1987]. *All in the Timing: Fourteen Plays*. New York: Vintage Books, 1995. 19–30.
20. Ibid., 21.
21. Ibid.
22. Ibid., 25.
23. Ibid., 22.
24. Ibid., 28.
25. Ibid., 23.
26. Ibid., 27.
27. Ibid., 28.
28. Ibid., 29.
29. Ibid., 26.
30. Ibid., 29.
31. Ibid., 30.

Chapter 23

1. Cronin, Doreen. *Click, Clack, Moo: Cows That Type*. New York: Simon & Schuster, 2000. 4.
2. <en.wikipedia.org/wiki/Principia_Discordia>
3. <www.principiadiscordia.com/book/55.php>
4. Cronin, Doreen. *Click, Clack, Moo*, 25.
5. Ibid., 6.
6. Ibid., 9.
7. Ibid., 10.
8. Ibid., 14, 16.
9. Ibid., 22.
10. Ibid., 23.
11. Ibid., 25.
12. Cronin, Doreen. *Duck for President*. New York: Simon & Schuster, 2004.

Chapter 24

1. Bender, Marilyn. "Think Tank: Heredity Factor Denied." *Globe and Mail* 19 June 1969, W8.

2. Borgese, Elisabeth Mann. *The Language Barrier: Beasts and Men*. New York: Holt, Rinehart and Winston, 1965.
3. Ibid.
4. <animal.discovery.com/news/briefs/20040607/dogspeak.html>
5. Gooderham, Mary. "Enlightened World View." *Globe and Mail* 3 March 1990, D1, D8.
6. Borgese, 39–40.
7. Ibid., 46.
8. Ibid.
9. Ibid., 47.
10. Ibid., 53.
11. Ibid., 49–50.
12. Ibid., 55.
13. Ibid., 60.
14. Ibid., 60, 63.
15. Ibid., 64.

Chapter 25

1. White, E. B. "Introduction." In Marquis, Don. *The Lives and Times of Archy & Mehitabel*. Illus. George Herriman. New York: Doubleday & Company, 1950. xvii–xxiv. xviii.
2. Marquis, Don. *The Lives and Times of Archy & Mehitabel*. Illus. George Herriman. New York: Doubleday & Company, 1950. 19.
3. Ibid., 20.
4. White, E. B. "Introduction." xviii.
5. Marquis, Don. "archy protests." *The Lives and Times of Archy & Mehitabel*. 202–3.
6. Ibid.
7. Marquis, Don. *The Lives and Times of Archy & Mehitabel*. 21.
8. White, E. B. "Introduction." xix.
9. Ibid., xvii.
10. Ibid., xxiii.
11. <www.donmarquis.com/don/index.html>
12. White, E. B. "Introduction." xx.
13. Ibid., xxi.
14. Marquis, Don. *The Lives and Times of Archy & Mehitabel*. 203.
15. Ibid., 204.
16. Ibid.

Chapter 26

1. Robbins, Tom. *Still Life with Woodpecker: A Sort of Love Story*. New York: Bantam Books, 1980. ix.
2. Ibid., 34.
3. Ibid., 204.
4. Ibid., 35.
5. Ibid., 36.
6. Ibid., 123.
7. Ibid.
8. Ibid., 204.
9. Ibid., 273–77.
10. Ibid., 36.

Part 5

Chapter 27

1. Ayre, John. *Northrop Frye*. Toronto: Random House, 1989. 51.
2. Ibid., 52.
3. Ibid., 53.
4. *Journal of Arts and Sciences*, 1823. Cited in Brauner, Ludwig. *Die Schreibmaschine in technischer, kultureller und wirtschaftlicher Bedetung*. Sammlung gemeinnütziger Vorträge, ed. Deutscher Verein zur Verbreitung gemeinnütziger Kenntnisse in Prag. Prague, 1925. 4. Qtd. in Kittler, 189.
5. Bliven, Bruce, Jr. *The Wonderful Writing Machine*. New York: Random House, 1954. 35.
6. Ibid., 14.
7. Ibid., 35.
8. <inventors.about.com/library/inventors/bltelegraph.htm>
9. Twain, Mark. "The First Writing-Machines." *The Complete Essays of Mark Twain*. Ed. Charles Nieder. Cambridge/New York: Da Capo Press, 2000. 324–26. 326.
10. Herkimer, 72; Bliven, 62.
11. Kittler, 192.
12. Ibid., 324, 326.
13. Ibid., 325, 326.
14. Ibid., 325.
15. Ibid., 326.

16. "Growth and Expertise of the Secretarial Profession is Closely Aligned with Office Equipment Technology." [Chart] Professional Secretaries International Organization. 1990. In Russo, 25.
17. Twain, Mark. "The First Writing-Machines." 324.
18. <www.infoplease.com/ce6/people/A0834880.html>
19. Twain, Mark. "The First Writing-Machines." 325.
20. Ibid., 326.
21. Ibid., 325.
22. Ibid.
23. Ibid., 326.
24. Ibid.
25. Herkimer, 73–74. Also reproduced in Russo, 20.
26. Ibid.
27. Ibid., 325.

Chapter 28

1. Foucault, Michel. "What Is an Author?" 133.
2. Bliven, Bruce, Jr., 111.
3. Russo, 25.
4. Russo, 25.
5. Herkimer, 111.
6. Ibid., 112.
7. Bliven, Bruce Jr., 112.
8. Qtd. in Bliven 112.
9. Ibid.
10. Beeching, 40.
11. Russo, 25.
12. Beeching, 40.
13. Yamada, Hisao. "Design Problems Associated with Japanese Keyboards." Ed. William E. Cooper, *Cognitive Aspects of Skilled Typewriting*. New York: Springer-Verlag, 1983. 305–407. 323.
14. Cooper, 21.
15. Ibid., 33.
16. Russo, 97.
17. Bliven, 114–15.
18. Ibid., 115.
19. Ibid., 116.
20. Bliven, 127.

21. Bliven, 126–27.
22. Bliven, 124–25.
23. Ibid.
24. Ibid., 124.
25. Ibid., 129.
26. <www.fivestarstaff.com/publication_typing.htm.>

Chapter 29

1. Kerouac, Jack. "[Here I Am at Last with a Typewriter]." *Atop an Underwood: Early Stories and Other Writings*. Ed. Paul Marion. New York: Viking Penguin, 1999. 130–31.
2. Marion, Paul. N. in Kerouac, Jack. "[Atop an Underwood: Introduction]." *Atop an Underwood: Early Stories and Other Writings*. Ed. Paul Marion. New York: Viking Penguin, 1999. 132–34. 133.
3. Kerouac, Jack. "[Atop an Underwood: Introduction]." *Atop an Underwood: Early Stories and Other Writings*. Ed. Paul Marion. New York: Viking Penguin, 1999. 132–34. 134.
4. Hunt, Tim. *Kerouac's Crooked Road: Development Of a Fiction*. [1981]. Berkeley: University of California Press, 1996. 101.
5. Ibid., 107.
6. Ibid., 110.
7. Kerouac, Jack. "The Origins of Joy in Poetry." [1958]. *Good Blonde & Others*. San Francisco: Grey Fox, 1993. 74.
8. Boon, Marcus. *The Road of Excess: A History of Writers on Drugs*. Cambridge/London: Harvard University Press, 2002. 197.
9. Hunt, Tim. *Kerouac's Crooked Road*, 110.
10. Nicoisia, Gerald. *Memory Babe: A Critical Biography of Jack Kerouac*. [1983]. Harmondsworth, Middlesex: Penguin Books, 1986. 33.
11. Ibid., 110.
12. Holmes, John Clellon. *Nothing More to Declare*. New York: E. P. Dutton & Co, 1967. 78. Qtd. in Hunt 110.
13. Burroughs, William S. Letter to Allen Ginsberg, 9 November 1948. Qtd. in Hunt 84. Hunt argues that Kerouac "must have" received a similar letter simultaneously because of his sudden interest in factualism, corresponding with the beginning of *On the Road*.
14. Albright, Alex. "Ammons, Kerouac and Their New Romantic Scrolls." *Complexities of Motion: New Essays On A. R. Ammons' Long Poems*.

Ed. Steven P. Schneider. Madison/Teaneck: Fairleigh Dickinson University Press, 1999. 83–110.

15. Holmes, John Clellon. *Nothing More to Declare*. 78–79. Qtd. in Boon 197.
16. Ann Charters, in Kerouac's letters. Kerouac, Jack. *Letters 1957–1969*. Ed. Ann Charters. New York: Viking Penguin, 1999. 461.
17. Giroux's memory. Hunt, Tim. *Kerouac's Crooked Road*, 112–13.
18. Ibid.
19. Boon, 197–98.
20. Kerouac, Jack. *Letters 1957–1969*. Ed. Ann Charters. New York: Viking Penguin, 1999. 395.
21. Kerouac, Jack. Preface. *Excerpts from Visions of Cody*. New York: New Directions, 1960. 5. Qtd. in Hunt 121.
22. Nicoisia, Gerald. *Memory Babe: A Critical Biography of Jack Kerouac*. 588.
23. See Capote, Truman. *Truman Capote: Conversations*. Ed. M. Thomas Inge. Jackson/London: University Press of Mississippi, 1987. 299; also Capote, Truman. *Conversations with Capote*. Ed. Lawrence Grobel. New York: NAL Books, 1985. 135.
24. Capote, Truman. *Conversations with Capote*. Ed. Lawrence Grobel. New York: NAL Books, 1985. 135.
25. Capote, Truman. *Truman Capote: Conversations*. Ed. M. Thomas Inge. Jackson/London: University Press of Mississippi, 1987. 298.
26. Mailer, Norman. "Of a Small and Modest Malignancy." *Esquire*. Qtd. in Capote, Truman. Conversations with Capote. Ed. Lawrence Grobel. New York: NAL Books, 1985. 32
27. Ibid., 198.

Chapter 30

1. Salthouse, Timothy. "Die Fertigkeit des Maschinenschreibens." *Spektrum dir Wissenschaft* 4: 94–100. 94–96. [Qtd. in Kittler 190.]
2. Virilio, Paul. *Speed and Politics*. [1977]. Trans. Mark Polizzotti. New York: Semiotext(e), 1986. 46.
3. Ibid., 47.
4. Ibid., 48–49.
5. Ibid., 18.
6. <www.auto-ordnance.com/ao_ao.html>
7. Ibid.
8. <en.wikipedia.org/wiki/Chicago_typewriter>

9. cf. Marshall McLuhan, *Understanding Media*, 228.
10. Bliven, 22–23.
11. Vonnegut, Kurt. *Mother Night*. [1961]. New York: Delta Books, 1996. 1.
12. Ibid., 2.
13. Virilio, *Speed and Politics*, 22.
14. <en.wikipedia.org/wiki/Samizdat>
15. <en.wikipedia.org/wiki/First_Department>
16. Vadislav, Jan. "All You Need Is a Typewriter." *Index on Censorship* 2 (1983): 33–35. 33.
17. Tomin, Zdena. "The Typewriters Hold the Fort." *Index On Censorship* 2 (1983): 28–30. 28.
18. Tomin, 28.
19. Vadislav, 34.
20. Ibid.
21. Ibid., 35.
22. Tomin, 28.
23. Ibid.
24. Sterling, Bruce. "Remarks at Computers, Freedom and Privacy Conference IV." Chicago, 26 March, 1994. <www.langston.com/Fun_People/1994/1994AIF.html>

Chapter 31

1. Beeching, 122–23.
2. Ibid., 124.
3. Ibid., 127.
4. Ibid.
5. Ibid., 124.

Epilogue

1. Ranaldo, Lee. "William S. Burroughs I-View." *A Burroughs Compendium: Calling the Toads*. Ed. Denis Mahoney, Richard L. Martin, and Ron Whitehead. Watch Hill: Ring Tarigh, 1998. 77–93. 92–93.

Part 6

1. See sites like The Best Case Scenario <www.thebestcasescenario.com> or the Mini-ITX Projects page <www.mini-itx.com/projects.asp>, or simply Google "case mods" for examples.

2. <www.mini-itx.com/projects/underwood/>
3. <www.mini-itx.com/projects/underwood/?page=3>
4. *The Onion* 39.41 <theonion.com/3941/>.
5. Fisher, Marshall Jon. "Memoria Ex Machina." *Harper's Magazine* 305.1831 (December 2002). 22–26. 26.
6. <www.csmonitor.com/2004/0803/p14s01-legn.html>
7. Ibid.
8. <ask.slashdot.org/article.pl?sid=04/08/04/1919229&tid=146&tid=126&tid=4>
9. *The Order of Things: An Archaeology of the Human Sciences*. [1966]. New York: Vintage Books, 1973. 50.
10. Williams, Robin. *The Mac Is Not a Typewriter*. 2nd ed. Berkeley: Peachpit Press, 2003. 13.
11. Ibid.
12. Ibid., 14.
13. See esp. Haraway, Donna. "A Cyborg Manifesto: Science, Technology, and Socialist-Feminism in the Late Twentieth Century." *Simian, Cyborgs and Women: The Reinvention of Nature*. New York: Routledge, 1991. 149–81.
14. Womack, Jack. *Ambient*. New York: Weidenfield & Nicolson, 1987. 50.
15. Vollmann, William T. *You Bright and Risen Angels*. London: Andre Deutsch, 1987. 15.
16. Barth, John. *Lost in the Funhouse: Fiction for Print, Tape, Live Voice*. [1968]. New York: Doubleday, 1988. 97.
17. Vollmann, 17.
18. De Landa, Manuel. *War in the Age of the Intelligent Machine*. New York: Zone Books, 1991. 130.
19. Ibid., 130–31.
20. McLuhan, Marshall. *Understanding Media: The Extensions of Man*. New York: Signet Books, 1964. 23.
21. De Landa, 148.
22. <en.wikipedia.org/wiki/Player_piano>
23. <www.sciencemuseum.org.uk/on-line/babbage/page3.asp>
24. <en.wikipedia.org/wiki/Alternating_current>
25. <en.wikipedia.org/wiki/Telephone>
26. <en.wikipedia.org/wiki/ARPANET>
27. Kahn, David. *The Codebreakers: The Story of Secret Writing*. New York: The Macmillan Company, 1967. 411.

28. Ibid., 411.
29. Ibid., 412–13.
30. Ibid., 411.
31. Ibid., 415.
32. Ibid., 420, 422.
33. Kittler, 242.
34. Ibid., 243.
35. For information on the principles of bazaar-style development, see Raymond, Eric S. "The Cathedral and the Bazaar." <www.catb.org/~esr/>.
36. Ellis, Warren, and Darick Robertson, illus. Nathan Eyring, colourist; Clem Robins and John Costanza, lettering. *Transmetropolitan*. Vol. 0: *Tales of Human Waste*. New York: Vertigo/DC Comics, 2004. 91.
37. Groening, Matt. "Akbar and Jeff's Piercing Hut." *Life in Hell*, 1997.
38. Ellis, Warren, and Darick Robertson, illus. Kim Demulder and Rodney Ramos, inkers. Nathan Eyring, colourist; Clem Robins, lettering. *Transmetropolitan*. Vol. 2: *Lust for Life*. New York: Vertigo/DC Comics, 1998. 105.
39. Ibid., 107–8.
40. Wolfe, Gene. "Appendix: A Note on the Translation." *The Shadow of the Torturer*. New York: Timescape Books, 1980. 275–76.
41. Ellis, Warren, and Darick Robertson, illus. Nathan Eyring, colourist; Clem Robins and John Costanza, lettering. *Transmetropolitan*. Vol. 0: *Tales of Human Waste*. New York: Vertigo/DC Comics, 2004. 47.
42. ———, illus. Kim Demulder and Rodney Ramos, inkers. Nathan Eyring, colourist; Clem Robins, lettering. *Transmetropolitan*. Vol. 2: *Lust for Life*. New York: Vertigo/DC Comics, 1998. 26.
43. ———, illus. Rodney Ramos, inker. Nathan Eyring, colourist; Clem Robins, lettering. *Transmetropolitan*. Vol. 10: *One More Time*. New York: Vertigo/DC Comics, 2004. 63.
44. Ibid., 128.
45. ———, illus. Keith Aiken, Jerome K. Moore, Ray Kryssing, Dick Giordano, inkers. Nathan Eyring, colour and separations; Clem Robins, letterer. *Transmetropolitan*. Vol 1: *Back on the Street*. New York: Vertigo/DC Comics, 1998. 46, 58, 60, 66.
46. ———, illus. Kim Demulder and Rodney Ramos, inkers. Nathan Eyring, colourist; Clem Robins, lettering. *Transmetropolitan*. Vol. 2: *Lust for Life*. New York: Vertigo/DC Comics, 1998. 4.

47. ———, illus. Keith Aiken, Jerome K. Moore, Ray Kryssing, Dick Giordano, inkers. Nathan Eyring, colour and separations; Clem Robins, letterer. *Transmetropolitan*. Vol. 1: *Back on the Street*. New York: Vertigo/DC Comics, 1998, back cover.
48. ———, illus. Keith Aiken, Jerome K. Moore, Ray Kryssing, Dick Giordano, inkers. Nathan Eyring, colour and separations; Clem Robins, letterer. *Transmetropolitan*. Vol. 1: *Back on the Street*. New York: Vertigo/DC Comics, 1998. 62.
49. ———, illus. Kim Demulder and Rodney Ramos, inkers. Nathan Eyring, colourist; Clem Robins, lettering. *Transmetropolitan*. Vol. 2: *Lust for Life*. New York: Vertigo/DC Comics, 1998. 112.
50. ———, illus. Keith Aiken, Jerome K. Moore, Ray Kryssing, Dick Giordano, inkers. Nathan Eyring, colour and separations; Clem Robins, letterer. *Transmetropolitan*. Vol. 1: *Back on the Street*. New York: Vertigo/DC Comics, 1998. 59; 62.
51. Ellis, Warren, and Darick Robertson, illus. Nathan Eyring, colourist; Clem Robins and John Costanza, lettering. *Transmetropolitan*. Vol. 0: *Tales of Human Waste*. New York: Vertigo/DC Comics, 2004. 68.
52. "Boyfriend Is a Virus." Ellis, Warren, and Darick Robertson, illus. Kim Demulder and Rodney Ramos, inkers. Nathan Eyring, colourist; Clem Robins, lettering. *Transmetropolitan*. Vol 2: *Lust for Life*. New York: Vertigo/DC Comics, 1998. 72–92.
53. <www.sarai.net/journal/03pdf/243_246_glovink.pdf>
54. Ellis, Warren, and Darick Robertson, illus. Keith Aiken, Jerome K. Moore, Ray Kryssing, Dick Giordano, inkers. Nathan Eyring, colour and separations; Clem Robins, letterer. *Transmetropolitan*. Vol. 1: *Back on the Street*. New York: Vertigo/DC Comics, 1998. 58.
55. Ibid.
56. Ibid.
57. Ellis, Warren, and Darick Robertson, illus. Rodney Ramos, inker. Nathan Eyring, colourist; Clem Robins, letterer. *Transmetropolitan*. Vol. 3: *Year of the Bastard*. New York: Vertigo/DC Comics, 1999. 116.
58. <www.cbsnews.com/stories/2004/09/20/politics/main644546.shtml>
59. <www.cbsnews.com/stories/2004/09/20/eveningnews/main644664.shtml>
60. Ibid.
61. <www.cbsnews.com/stories/2004/09/15/60II/main643768.shtml>

62. Baudrillard, Jean. Simulations. Trans. Paul Foss, Paul Patton, Philip Beitchman. New York: Semiotext(e).
63. <www.blogger.com>, <news.postnuke.com>, <www.movabletype.org/>
64. <shirky.com/writings/Weblogs_publishing.html>
65. <www.powerlineblog.com/archives/007760.php>
66. Ibid.
67. Ibid.
68. <www.dailykos.com/story/2004/9/10/34914/1603>
69. <www.washingtonpost.com/wp-dyn/articles/A9967-2004Sep9.html>; <www.washingtonpost.com/wpsrv/nation/daily/graphics/guard_091404.html>
70. <query.nytimes.com/gst/abstract.html?res=F70B14FD3E540C738DDDA00894DC404482>
71. <archive.salon.com/politics/war_room/archive.html?day=20040910>
72. <www.cbsnews.com/stories/2004/09/06/politics/main641481.shtml>
73. Bakhtin, Mikhail. *Problems of Dostoyevsky's Poetics*. Ed. and Trans. Caryl Emerson. Theory and History of Literature 8. Minneapolis: University of Minnesota Press, 1984. 16.
74. Ibid.
75. "CBS Fires 4 in Reporting of Bush National Guard Story (Update4)." Bloomberg. 10 January 2005. <www.bloomberg.com/apps/news?pid=10000103&sid=alDm3MtlxgqY&refer=us>
76. Graham, Rodney. *Rheinmetall/Victoria 8*. Silent film projection. 35 mm film, projector, looper. 10:50 minutes, continuous loop. 2003. *Rodney Graham: A Little Thought*. Art Gallery of Ontario: Toronto, 31 March–27 June 2004. See also <www.donaldyoung.com/graham/graham_1.html>.
77. *Rodney Graham: A Little Thought*. Exhibition Guide. Art Gallery of Ontario: Toronto, 31 March–27 June 2004. 2.
78. Ibid., 23.
79. Ibid.
80. "Strangled by an Intestine: An Interview with Guy Maddin." *Virus* 23 $ [third issue] (Spring 1992): 11–14. 12.
81. *Rodney Graham: A Little Thought*. Exhibition Guide. Art Gallery of Ontario: Toronto, 31 March–27 June 2004. 23.

Index